Praise for *Field Notes f[...]*

"Each chapter is part of a larger narrative, a loose travelogue that includes the Alaskan interior, Iceland and the Greenland ice sheet, but, more important, these narrative elements, while drawing us in, always keep a larger purpose in sight—to offer the clearest view yet of the biggest catastrophe we have ever faced."
—*Los Angeles Times*

"Comprehensive and succinct."—*New York Times*

"Elegant."—*Chicago Tribune*

"Kolbert weighs in with a slim but very powerful book."
—*Chicago Sun-Times*

"Essential reading. [Kolbert] is an accomplished writer whose prose is deceptively simple and whose meaning is always clear. Climate change is complex stuff, but she deftly distills the brew to clarity."—*Minneapolis Star Tribune*

"Gripping, well-written."—*Seattle Times*

"Short, readable and scrupulously objective . . . Kolbert's central achievement is to have examined the complex panoply of data surrounding the science of climatology and to have presented it in prose as elegant as the facts themselves are stark."—*Times Literary Supplement*

"Very powerful . . . important."—*Hartford Courant*

"Enlightening."—*Oregonian*

"*Field Notes* is a wonderful read, a superbly crafted, diligently compressed vision of a world spiralling towards destruction. It should be a wake-up call to the world."—*Observer* **(London)**

"Kolbert is like Matisse, painting an evocative picture with a few deft strokes . . . I recommend [*Field Notes*] to anyone who has a concern for the fate of our planet."—*American Scientist*

"Kolbert, a staff writer for *The New Yorker*, surveyed the world's leading climate scientists, and discovered an alarming unanimity to their message: The world needs to wake up, and fast." —*Wired*

"Let's hope this powerful, clear and important book is not just lightly compared to *Silent Spring*. Let's hope it is this era's galvanizing text."—*Scientific American*

"An elegant ride through the confusing world of climate science. Kolbert takes a John McPhee–style ramble across the world: In Greenland, Iceland, and poor little Shishmaref, she sees the effects of warming firsthand. In Washington, D.C., the former *New York Times* political reporter puts her Beltway savvy to use, revealing that the most climate-change skepticism originates in the deep pockets of oil and coal companies."—*Outside*

"Keenly observed and deeply memorable . . . the picture [Kolbert] draws is compelling—and very scary."—*Seed*

"An extraordinary piece of reporting."—**Grist.org**

"Kolbert forces us to ponder a tragic disconnect: Politicians, the group best positioned to do something about the scientists' warnings, turn out to be the group that's most adamantly ignoring those warnings."—*Austin American-Statesman*

"[Kolbert] traverses the globe to observe first hand the jaw-dropping results of climate changes . . . species going extinct, polar caps melting, water rising in the Netherlands or butterflies in England migrating to higher elevations than ever before." —*Contra Costa Times*

The Prophet of Love
The Sixth Extinction

FIELD NOTES

from a

CATASTROPHE

MAN, NATURE, AND CLIMATE CHANGE

Elizabeth Kolbert

BLOOMSBURY

NEW YORK · LONDON · OXFORD · NEW DELHI · SYDNEY

Bloomsbury USA
An imprint of Bloomsbury Publishing Plc

1385 Broadway   50 Bedford Square
New York   London
NY 10018   WC1B 3DP
USA   UK

www.bloomsbury.com

BLOOMSBURY and the Diana logo are trademarks of Bloomsbury Publishing Plc

First published 2006
This paperback edition published 2015
Updated with new material in 2015

No responsibility for loss caused to any individual or organization acting on or
refraining from action as a result of the material in this publication can be
accepted by Bloomsbury or the author.

ISBN: TPB: 978-1-62040-988-6
        ePub: 978-1-62040-989-3

Library of Congress Cataloging-in-Publication Data

Kolbert, Elizabeth.
Field notes from a catastrophe : man, nature, and climate change / Elizabeth
Kolbert.—1st U.S. ed.
p. cm.
Includes bibliographical references and index.
ISBN 978-1-62040-988-6 (paperback)
1. Global warming. 2. Global temperature changes. 3. Global environmental
change. I. Title.
QC981.8.G56K655 2006
363.738'74—dc22
2005030972

7 9 10 8 6

Typeset by Hewer Text UK Ltd, Edinburgh
Printed and bound in the U.S.A. by Berryville Graphics, Berryville, Virginia

To find out more about our authors and books visit www.bloomsbury.com.
Here you will find extracts, author interviews, details of forthcoming events,
and the option to sign up for our newsletters.

Bloomsbury books may be purchased for business or promotional use. For information
on bulk purchases please contact Macmillan Corporate and Premium Sales Department
at specialmarkets@macmillan.com.

To my boys

AUTHOR'S NOTE

The language of science is metric; however, most British and American readers speak—and think—in units like feet, miles, and degrees Fahrenheit. I have used English units where practical and metric units where it seemed clearly more appropriate. For instance, the standard measure of carbon emissions is metric tons. A metric ton weighs 2,205 pounds.

CONTENTS

PREFACE

THERE ISN'T MUCH to do at the Hotel Arctic except watch the icebergs flow by. The hotel is located in the town of Ilulissat, on the west coast of Greenland, four degrees north of the Arctic Circle. The icebergs originate some fifty miles away, at the end of a long and fast-moving ice stream known as the Jakobshavn Isbrae. They drift down a fjord and through a wide-mouthed bay, and, if they last long enough, end up in the North Atlantic. (It is likely that the iceberg encountered by the *Titanic* followed this route.)

To the tourists who visit the Hotel Arctic, the icebergs are a thrilling sight: beautiful and terrible in equal measure. They are a reminder of the immensity of nature and the smallness of man. To the people who spend more time in Ilulissat—native Greenlanders, European tour guides, American scientists—the icebergs have come to acquire a different significance. Since the late 1990s, the Jakobshavn Isbrae has doubled its speed. In the process, the height of the ice stream has been dropping by up to fifty feet a year and the calving front has retreated by several miles. What locals now notice about the icebergs is not their power or immensity—though they are still powerful and immense—but a disquieting diminishment.

"You don't get the big icebergs anymore," an Ilulissat town councilman named Jeremias Jensen told me. We were having coffee on a late-spring afternoon in the Hotel Arctic lobby. Outside, it was foggy and the icebergs seemed to be rising up out of the mist. "It's very strange the last few years; you can see a lot of strange changes."

This is a book about watching the world change. It grew out of three articles that I wrote for the *New Yorker* magazine, which ran in the spring of 2005, and its goal remains much the same as that of the original series: to convey, as vividly as possible, the reality of global warming. The opening chapters are set near or above the Arctic Circle—in Deadhorse, Alaska; in the countryside outside of Reykjavík; at Swiss Camp, a research station on the Greenland ice sheet. I went to these particular places for all the usual journalistic reasons—because someone invited me to tag along on an expedition, because someone let me hitch a ride on a helicopter, because someone sounded interesting over the telephone. The same is true of the choices that were made in subsequent chapters, whether it was a decision to track butterflies in northern England or to visit floating houses in the Netherlands. Such is the impact of global warming that I could have gone to hundreds if not thousands of other places—from Siberia to the Austrian Alps to the Great Barrier Reef to the South African fynbos—to document its effects. These alternate choices would have resulted in an account very different in its details, but not in its conclusions.

Humans aren't the first species to alter the atmosphere; that distinction belongs to early bacteria, which, some two billion years ago, invented photosynthesis. But we are the first species to be in a position to understand what we are doing. Computer models of the earth's climate suggest that a critical threshold is approaching. Crossing over it will be easy, crossing back quite likely impossible. The second part of this book explores the complicated relationship between the science and the politics of global warming, between what we know and what we refuse to know.

My hope is that this book will be read by everyone, by which I mean not only those who follow the latest news about the climate but also those who prefer to skip over it. For better or (mostly) for worse, global warming is all about scale, and the sheer number of figures involved can be daunting. I've tried to offer what is essential without oversimplifying. Similarly, I have tried to keep the discussion of scientific theory to a minimum while offering a full-enough account to convey what is truly at stake.

Part I

NATURE

Chapter 1

SHISHMAREF, ALASKA

THE ALASKAN VILLAGE of Shishmaref sits on an
island known as Sarichef, five miles off the coast of the
Seward Peninsula. Sarichef is a small island—no more than
a quarter of a mile across and two and a half miles long—
and Shishmaref is basically the only thing on it. To the
north is the Chukchi Sea, and in every other direction lies
the Bering Land Bridge National Preserve, which probably
ranks as one of the least visited national parks in the
country. During the last ice age, the land bridge—exposed
by a drop in sea levels of more than three hundred feet—
grew to be nearly a thousand miles wide. The preserve
occupies that part of it which, after more than ten thousand
years of warmth, still remains above water.

Shishmaref (population 591) is an Inupiat village, and it
has been inhabited, at least on a seasonal basis, for several
centuries. As in many native villages in Alaska, life there
combines—often disconcertingly—the very ancient and
the totally modern. Almost everyone in Shishmaref still
lives off subsistence hunting, primarily for bearded seals but
also for walrus, moose, rabbits, and migrating birds. When

I visited the village one day in April, the spring thaw was under way, and the seal-hunting season was about to begin. (Wandering around, I almost tripped over the remnants of the previous year's catch emerging from storage under the snow.) At noon, the village's transportation planner, Tony Weyiouanna, invited me to his house for lunch. In the living room, an enormous television set tuned to the local public-access station was playing a rock soundtrack. Messages like "Happy Birthday to the following elders . . ." kept scrolling across the screen.

Traditionally, the men in Shishmaref hunted for seals by driving out over the sea ice with dogsleds or, more recently, on snowmobiles. After they hauled the seals back to the village, the women would skin and cure them, a process that takes several weeks. In the early 1990s, the hunters began to notice that the sea ice was changing. (Although the claim that the Eskimos have hundreds of words for snow is an exaggeration, the Inupiat make distinctions among many different types of ice, including *sikuliaq*, "young ice," *sarri*, "pack ice," and *tuvaq*, "land-locked ice.") The ice was starting to form later in the fall, and also to break up earlier in the spring. Once, it had been possible to drive out twenty miles; now, by the time the seals arrived, the ice was mushy half that distance from shore. Weyiouanna described it as having the consistency of a "slush puppy." When you encounter it, he said, "your hair starts sticking up. Your eyes are wide open. You can't even blink." It became too dangerous to hunt using snowmobiles, and the men switched to boats.

Soon, the changes in the sea ice brought other problems. At its highest point, Shishmaref is only twenty-two feet above sea level, and the houses, most of which were built by the U.S. government, are small, boxy, and not particularly sturdy-looking. When the Chukchi Sea froze early, the layer of ice protected the village, the way a tarp prevents a swimming pool from getting roiled by the wind. When the sea started to freeze later, Shishmaref became more vulnerable to storm surges. A storm in October 1997 scoured away a hundred-and-twenty-five-foot-wide strip from the town's northern edge; several houses were destroyed, and more than a dozen had to be relocated. During another storm, in October 2001, the village was threatened by twelve-foot waves. In the summer of 2002, residents of Shishmaref voted, a hundred and sixty-one to twenty, to move the entire village to the mainland. In 2004, the U.S. Army Corps of Engineers completed a survey of possible sites. Most of the spots that are being considered for a new village are in areas nearly as remote as Sarichef, with no roads or nearby cities or even settlements. It is estimated that a full relocation would cost the U.S. government $180 million.

People I spoke to in Shishmaref expressed divided emotions about the proposed move. Some worried that, by leaving the tiny island, they would give up their connection to the sea and become lost. "It makes me feel lonely," one woman said. Others seemed excited by the prospect of gaining certain conveniences, like running water, that Shishmaref lacks. Everyone seemed to agree, though, that the village's situation, already dire, was only going to get worse.

Morris Kiyutelluk, who is sixty-five, has lived in Shishmaref almost all his life. (His last name, he told me, means "without a wooden spoon.") I spoke to him while I was hanging around the basement of the village church, which also serves as the unofficial headquarters for a group called the Shishmaref Erosion and Relocation Coalition. "The first time I heard about global warming, I thought, I don't believe those Japanese," Kiyutelluk told me. "Well, they had some good scientists, and it's become true."

The National Academy of Sciences undertook its first major study of global warming in 1979. At that point, climate modeling was still in its infancy, and only a few groups, one led by Syukuro Manabe at the National Oceanic and Atmospheric Administration and another by James Hansen at NASA's Goddard Institute for Space Studies, had considered in any detail the effects of adding carbon dioxide to the atmosphere. Still, the results of their work were alarming enough that President Jimmy Carter called on the academy to investigate. A nine-member panel was appointed. It was led by the distinguished meteorologist Jule Charney, of MIT, who, in the 1940s, had been the first meteorologist to demonstrate that numerical weather forecasting was feasible.

The Ad Hoc Study Group on Carbon Dioxide and Climate, or the Charney panel, as it became known, met for five days at the National Academy of Sciences' summer study center, in Woods Hole, Massachusetts. Its conclusions were unequivocal. Panel members had looked for flaws in the modelers' work but had been unable to find any. "If carbon dioxide continues to increase, the study group finds no reason to doubt that climate changes will result and no reason to believe that these changes will be negligible," the scientists wrote. For a doubling of CO_2 from preindustrial levels, they put the likely global temperature rise at between two and a half and eight degrees Fahrenheit. The panel members weren't sure how long it would take for changes already set in motion to become manifest, mainly because the climate system has a built-in time delay. The effect of adding CO_2 to the atmosphere is to throw the earth out of "energy balance." In order for balance to be restored—as, according to the laws of physics, it eventually must be—the entire planet has to heat up, including the oceans, a process, the Charney panel noted, that could take "several decades." Thus, what might seem like the most conservative approach—waiting for evidence of warming to make sure the models were accurate—actually amounted to the riskiest possible strategy: "We may not be given a warning until the CO_2 loading is such that an appreciable climate change is inevitable."

It is now more than twenty-five years since the Charney panel issued its report, and, in that period, Americans have been alerted to the dangers of global warming so many times that reproducing even a small fraction of these warnings would fill several volumes;

indeed, entire books have been written just on the history of efforts to draw attention to the problem. (Since the Charney report, the National Academy of Sciences alone has produced nearly two hundred more studies on the subject, including, to name just a few, "Radiative Forcing of Climate Change," "Understanding Climate Change Feedbacks," and "Policy Implications of Greenhouse Warming.") During this same period, worldwide carbon-dioxide emissions have continued to increase, from five billion to seven billion metric tons a year, and the earth's temperature, much as predicted by Manabe's and Hansen's models, has steadily risen. The year 1990 was the warmest year on record until 1991, which was equally hot. Almost every subsequent year has been warmer still. As of this writing, 1998 ranks as the hottest year since the instrumental temperature record began, but it is closely followed by 2002 and 2003, which are tied for second; 2001, which is third; and 2004, which is fourth. Since climate is innately changeable, it's difficult to say when, exactly, in this sequence natural variation could be ruled out as the sole cause. The American Geophysical Union, one of the nation's largest and most respected scientific organizations, decided in 2003 that the matter had been settled. At the group's annual meeting that year, it issued a consensus statement declaring, "Natural influences cannot explain the rapid increase in global near-surface temperatures." As best as can be determined, the world is now warmer than it has been at any point in the last two millennia, and, if current trends continue, by the end of the century it will likely be hotter than at any point in the last two million years.

In the same way that global warming has gradually ceased to be merely a theory, so, too, its impacts are no longer just hypothetical. Nearly every major glacier in the world is shrinking; those in Glacier National Park are retreating so quickly it has been estimated that they will vanish entirely by 2030. The oceans are becoming not just warmer but more acidic; the difference between daytime and nighttime temperatures is diminishing; animals are shifting their ranges poleward; and plants are blooming days, and in some cases weeks, earlier than they used to. These are the warning signs that the Charney panel cautioned against waiting for, and while in many parts of the globe they are still subtle enough to be overlooked, in others they can no longer be ignored. As it happens, the most dramatic changes are occurring in those places, like Shishmaref, where the fewest people tend to live. This disproportionate effect of global warming in the far north was also predicted by early climate models, which forecast, in column after column of FORTRAN-generated figures, what today can be measured and observed directly: the Arctic is melting.

Most of the land in the Arctic, and nearly a quarter of all the land in the Northern Hemisphere—some five and a half billion acres—is underlaid by zones of permafrost. A few months after I visited Shishmaref, I went back to Alaska to take a trip through the interior of the state with

Vladimir Romanovsky, a geophysicist and permafrost expert. I flew into Fairbanks—Romanovsky teaches at the University of Alaska, which has its main campus there—and when I arrived, the whole city was enveloped in a dense haze that looked like fog but smelled like burning rubber. People kept telling me that I was lucky I hadn't come a couple of weeks earlier, when it had been much worse. "Even the dogs were wearing masks," one woman I met said. I must have smiled. "I am not joking," she told me.

Fairbanks, Alaska's second-largest city, is surrounded on all sides by forest, and virtually every summer lightning sets off fires in these forests, which fill the air with smoke for a few days or, in bad years, weeks. In the summer of 2004, the fires started early, in June, and were still burning two and a half months later; by the time of my visit, in late August, a record 6.3 million acres—an area roughly the size of New Hampshire—had been incinerated. The severity of the fires was clearly linked to the weather, which had been exceptionally hot and dry; the average summertime temperature in Fairbanks was the highest on record, and the amount of rainfall was the third lowest.

On my second day in Fairbanks, Romanovsky picked me up at my hotel for an underground tour of the city. Like most permafrost experts, he is from Russia. (The Soviets more or less invented the study of permafrost when they decided to build their gulags in Siberia.) A broad man with shaggy brown hair and a square jaw, Romanovsky as a student had had to choose between playing professional hockey and becoming a geophysicist. He had opted for the latter, he told me, because "I was little bit better scientist than hockey player." He went on to earn two master's degrees and two Ph.D.s. Romanovsky came to get me at ten A.M.; owing to all the smoke, it looked like dawn.

Any piece of ground that has remained frozen for at least two years is, by definition, permafrost. In some places, like eastern Siberia, permafrost runs nearly a mile deep; in Alaska, it varies from a couple of hundred feet to a couple of thousand feet deep. Fairbanks, which is just below the Arctic Circle, is situated in a region of discontinuous permafrost, meaning that the city is pocked with regions of frozen ground. One of the first stops on Romanovsky's tour was a hole that had opened up in a patch of permafrost not far from his house. It was about six feet wide and five feet deep. Nearby were the outlines of other, even bigger holes, which, Romanovsky told me, had been filled with gravel by the local public-works department. The holes, known as thermokarsts, had appeared suddenly when the permafrost gave way, like a rotting floorboard. (The technical term for thawed permafrost is "talik," from a Russian word meaning "not frozen.") Across the road, Romanovsky pointed out a long trench running into the woods. The trench, he explained, had been formed when a wedge of underground ice had melted. The spruce trees that had been growing next to it, or perhaps on top of it, were now listing at odd angles, as if in a gale. Locally, such trees are called "drunken." A few of the spruces had fallen over. "These are very drunk," Romanovsky said.

In Alaska, the ground is riddled with ice wedges that were created during the last glaciation, when the cold earth cracked and the cracks filled with water. The wedges, which can be dozens or even hundreds of feet deep, tended to form in networks, so when they melt, they leave behind connecting diamond- or hexagon-shaped depressions. A few blocks beyond the drunken forest, we came to a house where the front yard showed clear signs of ice-wedge melt-off. The owner, trying to make the best of things, had turned the yard into a miniature-golf course. Around the corner, Romanovsky pointed out a house—no longer occupied—that basically had split in two; the main part was leaning to the right and the garage toward the left. The house had been built in the sixties or early seventies; it had survived until almost a decade ago, when the permafrost under it started to degrade. Romanovsky's mother-in-law used to own two houses on the same block. He had urged her to sell them both. He pointed out one, now under new ownership; its roof had developed an ominous-looking ripple. (When Romanovsky went to buy his own house, he looked only in permafrost-free areas.)

"Ten years ago, nobody cared about permafrost," he told me. "Now everybody wants to know." Measurements that Romanovsky and his colleagues at the University of Alaska have made around Fairbanks show that the temperature of the permafrost in many places has risen to the point where it is now less than one degree below freezing. In places where the permafrost has been disturbed, by roads or houses or lawns, much of it is already thawing. Romanovsky has also been monitoring the permafrost on the North Slope and has found that there, too, are regions where the permafrost is very nearly thirty-two degrees Fahrenheit. While thermokarsts in the roadbeds and talik under the basement are the sort of problems that really only affect the people right near—or above—them, warming permafrost is significant in ways that go far beyond local real estate losses. For one thing, permafrost represents a unique record of long-term temperature trends. For another, it acts, in effect, as a repository for greenhouse gases. As the climate warms, there is a good chance that these gases will be released into the atmosphere, further contributing to global warming. Although the age of permafrost is difficult to determine, Romanovsky estimates that most of it in Alaska probably dates back to the beginning of the last glacial cycle. This means that if it thaws, it will be doing so for the first time in more than a hundred and twenty thousand years. "It's really a very interesting time," Romanovsky told me.

The next morning, Romanovsky picked me up at seven. We were going to drive from Fairbanks nearly five hundred miles north to the town of Deadhorse, on Prudhoe Bay. Romanovsky makes the trip at least once a year, to collect data from the many electronic monitoring stations he has set up. Since the way was largely unpaved, he had rented a truck for the occasion. Its windshield was cracked in several places. When I suggested this could be a problem, Romanovsky assured me that it was "typical

Alaska." For provisions, he had brought along an oversize bag of Tostitos.

The road that we traveled along—the Dalton Highway—had been built for Alaskan oil, and the pipeline followed it, sometimes to the left, sometimes to the right. (Because of the permafrost, the pipeline runs mostly aboveground, on pilings that contain ammonia, which acts as a refrigerant). Trucks kept passing us, some with severed caribou heads strapped to their roofs, others belonging to the Alyeska Pipeline Service Company. The Alyeska trucks were painted with the disconcerting motto "Nobody Gets Hurt." About two hours outside Fairbanks, we started to pass through tracts of forest that had recently burned, then tracts that were still smoldering, and, finally, tracts that were still, intermittently, in flames. The scene was part Dante, part *Apocalypse Now*. We crawled along through the smoke. After another few hours, we reached Coldfoot, named, supposedly, for some gold prospectors who arrived at the spot in 1900, then got "cold feet" and turned around. We stopped to have lunch at a truck stop, which made up pretty much the entire town. Just beyond Coldfoot, we passed the tree line. An evergreen was marked with a plaque that read "Farthest North Spruce Tree on the Alaska Pipeline: Do Not Cut." Predictably, someone had taken a knife to it. A deep gouge around the trunk was bound with duct tape. "I think it will die," Romanovsky told me.

Finally, at around five P.M., we reached the turnoff for the first monitoring station. By now we were traveling along the edge of the Brooks Range and the mountains were purple in the afternoon light. Because one of Romanovsky's colleagues had nursed dreams—never realized—of traveling to the station by plane, it was situated near a small airstrip, on the far side of a quickly flowing river. We pulled on rubber boots and forded the river, which, owing to the lack of rain, was running low. The site consisted of a few posts sunk into the tundra, a solar panel, a two-hundred-foot-deep borehole with heavy-gauge wire sticking out of it, and a white container, resembling an ice chest, that held computer equipment. The solar panel, which the previous summer had been mounted a few feet off the ground, was now resting on the scrub. At first, Romanovsky speculated that this was a result of vandalism, but after inspecting things more closely, he decided that it was the work of a bear. While he hooked up a laptop computer to one of the monitors inside the white container, my job was to keep an eye out for wildlife.

For the same reason that it is sweaty in a coal mine—heat flux from the center of the earth—permafrost gets warmer the farther down you go. Under equilibrium conditions—which is to say, when the climate is stable—the very warmest temperatures in a borehole will be found at the bottom and temperatures will decrease steadily as you go higher. In these circumstances, the lowest temperature will be found at the permafrost's surface, so that, plotted on a graph, the results will be a tilted line. In recent decades, though, the temperature profile of

Alaska's permafrost has drooped. Now, instead of a straight line, what you get is shaped more like a sickle. The permafrost is still warmest at the very bottom, but instead of being coldest at the top, it is coldest somewhere in the middle, and warmer again toward the surface. This is a sign—and an unambiguous one—that the climate is heating up.

"It's very difficult to look at trends in air temperature, because it's so variable," Romanovsky explained after we were back in the truck, bouncing along toward Deadhorse. It turned out that he had brought the Tostitos to stave off not hunger but fatigue—the crunching, he said, kept him awake—and by now the enormous bag was more than half empty. "So one year you have around Fairbanks a mean annual temperature of zero"—thirty-two degrees Fahrenheit—"and you say, 'Oh yeah, it's warming,' and other years you have mean annual temperature of minus six"—twenty-one degrees Fahrenheit—"and everybody says, 'Where? Where is your global warming?' In the air temperature, the signal is very small compared to noise. What permafrost does is it works as low-pass filter. That's why we can see trends much easier in permafrost temperatures than we can see them in atmosphere." In most parts of Alaska, the permafrost has warmed by three degrees since the early 1980s. In some parts of the state, it has warmed by nearly six degrees.

When you walk around in the Arctic, you are stepping not on permafrost but on something called the "active layer."

The active layer, which can be anywhere from a few inches to a few feet deep, freezes in the winter but thaws over the summer, and it is what supports the growth of plants— large spruce trees in places where conditions are favorable enough and, where they aren't, shrubs and, finally, just lichen. Life in the active layer proceeds much as it does in more temperate regions, with one critical difference. Temperatures are so low that when trees and grasses die they do not fully decompose. New plants grow on top of the half-rotted old ones, and when these plants die the same thing happens all over again. Eventually, through a process known as cryoturbation, organic matter is pushed down beneath the active layer into the permafrost, where it can sit for thousands of years in a botanical version of suspended animation. (In Fairbanks, grass that is still green has been found in permafrost dating back to the middle of the last ice age.) This is the reason that permafrost, much like a peat bog or, for that matter, a coal deposit, acts as a storage unit for accumulated carbon.

One of the risks of rising temperatures is that the storage process can start to run in reverse. Under the right conditions, organic material that has been frozen for millennia will begin to break down, giving off carbon dioxide or methane, which is an even more powerful (though more short-lived) greenhouse gas. In parts of the Arctic, this process is already under way. Researchers in Sweden, for example, have been measuring the methane output of a bog known as the Stordalen mire, near the town of Abisko, nine hundred miles north of Stockholm, for almost thirty-five years. As the permafrost in the area has warmed, methane releases have increased, in some spots by as much as 60 percent. Thawing permafrost could make the active layer more hospitable to plants, which are a sink for carbon. Even this, though, wouldn't be enough to offset the release of greenhouse gases. No one knows exactly how much carbon is stored in the world's permafrost, but estimates run as high as 450 billion metric tons.

"It's like ready-use mix—just a little heat, and it will start cooking," Romanovsky told me. It was the day after we had arrived in Deadhorse, and we were driving through a steady drizzle out to another monitoring site. "I think it's just a time bomb, just waiting for a little warmer conditions." Romanovsky was wearing a rain suit over his canvas work clothes. I put on a rain suit that he had brought along for me. He pulled a tarp out of the back of the truck.

Whenever he has had funding, Romanovsky has added new monitoring sites to his network. There are now sixty of them, and while we were on the North Slope he spent all day and also part of the night—it stayed light until nearly eleven—rushing from one to the next. At each site, the routine was more or less the same. First, Romanovsky would hook up his computer to the data logger, which had been recording permafrost temperatures on an hourly basis since the previous summer. When it was raining, Romanovsky would perform this first step hunched under the tarp. Then he would take out a metal probe shaped like a "T" and poke it into the ground at regular intervals, measuring the depth of the active layer. The probe was a meter long, which, it turned out, was no longer quite long enough. The summer had been so warm that almost everywhere the active layer had grown deeper, in some spots by just a few centimeters, in other spots by more than that. In places where the active layer was particularly deep, Romanovsky had had to work out a new way of measuring it using the probe and a wooden ruler. (I helped out by recording the results of this exercise in his waterproof field notebook.) Eventually, he explained, the heat that had gone into increasing the depth of the active layer would work its way downward, bringing the permafrost that much closer to the thawing point. "Come back next year," he advised me.

On the last day I spent on the North Slope, a friend of Romanovsky's, Nicolai Panikov, a microbiologist at the Stevens Institute of Technology, in New Jersey, arrived. He was planning on collecting cold-loving microorganisms known as psychrophiles, which he would take back to New Jersey to study. Panikov's goal was to determine whether the organisms could have functioned in the sort of conditions that, it is believed, were once found on Mars. He told me that he was quite convinced that Martian life existed—or, at least, had existed. Romanovsky expressed his opinion on this by rolling his eyes; nevertheless, he had agreed to help Panikov dig up some permafrost.

That same day, I flew with Romanovsky by helicopter to a small island in the Arctic Ocean, where he had set up yet another monitoring site. The island, just north of the seventieth parallel, was a bleak expanse of mud dotted with little clumps of yellowing vegetation. It was filled with ice wedges that were starting to melt, creating a network of polygonal depressions. The weather was cold and wet, so while Romanovsky hunched under his tarp I stayed in the helicopter and chatted with the pilot. He had lived in Alaska since 1967. "It's definitely gotten warmer since I've been here," he told me. "I have really noticed that."

When Romanovsky emerged, we took a walk around the island. Apparently, in the spring it had been a nesting site for birds, because everywhere we went there were bits of eggshell and piles of droppings. The island was only about ten feet above sea level, and at the edges it dropped off sharply into the water. Romanovsky pointed out a spot along the shore where the previous summer a series of ice wedges had been exposed. They had since melted, and the ground behind them had given way in a cascade of black mud. In a few years, he said, he expected more ice wedges would be exposed, and then these would melt, causing further erosion. Although the process was different in its mechanics from what was going on in Shishmaref, it had much the same cause and, according to Romanovsky, was likely to have the same result. "Another disappearing island," he said, gesturing toward some freshly exposed bluffs. "It's moving very, very fast."

On September 18, 1997, the *Des Groseilliers*, a three-hundred-and-eighteen-foot-long icebreaker with a bright-red hull, set out from the town of Tuktoyaktuk, on the

Beaufort Sea, and headed north under overcast skies. Normally, the *Des Groseilliers*, which is based in Québec City, is used by the Canadian Coast Guard, but for this particular journey it was carrying a group of American geophysicists, who were planning to jam it into an ice floe. The scientists were hoping to conduct a series of experiments as they and the ship and the ice floe all drifted, as one, around the Arctic Ocean. The expedition had taken several years to prepare for, and during the planning phase its organizers had carefully consulted the findings of a previous Arctic expedition, which had taken place back in 1975. The researchers aboard the *Des Groseilliers* were aware that the Arctic sea ice was retreating; that was, in fact, precisely the phenomenon they were hoping to study. Still, they were caught off guard. Based on the data from the 1975 expedition, they had decided to look for a floe averaging nine feet thick. When they reached the area where they planned to overwinter—at seventy-five degrees north latitude—not only were there no floes nine feet thick, there were barely any that reached six feet. One of the scientists on board recalled the reaction on the *Des Groseilliers* this way: "It was like 'Here we are, all dressed up and nowhere to go.' We imagined calling the sponsors at the National Science Foundation and saying, 'Well, you know, we can't find any ice.'"

Sea ice in the Arctic comes in two varieties. There is seasonal ice, which forms in the winter and then melts in the summer, and perennial ice, which persists year-round. To the untrained eye, all of it looks pretty much the same, but by licking it you can get a good idea of how long a particular piece has been floating around. When ice begins to form in seawater, it forces out the salt, which has no place in the crystal structure. As the ice thickens, the rejected salt collects in tiny pockets of brine too highly concentrated to freeze. If you suck on a piece of first-year ice, it will taste salty. Eventually, if the ice stays frozen long enough, these pockets of brine drain out through fine, veinlike channels, and the ice becomes fresher. Multiyear ice is so fresh that if you melt it, you can drink it.

The most precise measurements of Arctic sea ice have been made by NASA, using satellites equipped with microwave sensors. In 1979, the satellite data show, perennial sea ice covered 1.7 billion acres, or an area nearly the size of the continental United States. The ice's extent varies from year to year, but since then the overall trend has been strongly downward. The losses have been particularly great in the Beaufort and Chukchi Seas, and also considerable in the Siberian and Laptev Seas. During this same period, an atmospheric circulation pattern known as the Arctic Oscillation has mostly been in what climatologists call a "positive" mode. The positive Arctic Oscillation is marked by low pressure over the Arctic Ocean, and it tends to produce strong winds and higher temperatures in the far north. No one really knows whether the recent behavior of the Arctic Oscillation is independent of global warming or a product of it. By now, though, the perennial sea ice has shrunk by roughly 250 million acres, an area the size of New York, Georgia, and Texas combined. According to mathematical models, even the extended period of a positive Arctic Oscillation can account for only part of this loss.

At the time the *Des Groseilliers* set off, little information on trends in sea-ice depth was available. A few years later, a limited amount of data on this topic—gathered, for rather different purposes, by nuclear submarines—was declassified. It showed that between the 1960s and the 1990s, sea-ice depth in a large section of the Arctic Ocean declined by nearly 40 percent.

Eventually, the researchers on board the *Des Groseilliers* decided that they would just have to settle for the best ice floe they could find. They picked one that stretched over some thirty square miles. In some spots it was six feet thick, in some spots just three. Tents were set up on the floe to house experiments, and a safety protocol was established: anyone venturing out onto the ice had to travel with a buddy and a radio. (Many also carried a gun, in case of polar-bear problems.) Some of the scientists speculated that, since the ice was abnormally thin, it would grow thicker during the expedition. Just the opposite turned out to be the case. The *Des Groseilliers* spent twelve months frozen into the floe, and, during that time, it drifted some three hundred miles north. Nevertheless, at the end of the year, the average thickness of the ice had declined, in some spots by as much as a third. By August 1998, so many of the scientists had fallen through that a new requirement was added to the protocol: anyone who set foot off the ship had to wear a life jacket.

★   ★   ★

OBSERVED SEA ICE—SEPTEMBER 1979

OBSERVED SEA ICE—SEPTEMBER 2005

The extent of the Arctic's perennial sea ice has declined dramatically in recent years. Credit: F. Fetterer and K. Knowles, Sea Ice Index, National Snow and Ice Data Center.

Donald Perovich has studied sea ice for thirty years, and on a rainy day not long after I got back from Deadhorse, I went to visit him at his office in Hanover, New Hampshire. Perovich works for the Cold Regions Research and Engineering Laboratory, or CRREL (pronounced "crell"). CRREL is a division of the U.S. Army that was established in 1961 in anticipation of a very cold war. (The assumption was that if the Soviets invaded, they would probably do so from the north.) He is a tall man with black hair, very black eyebrows, and an earnest manner. His office is decorated with photographs from the *Des Groseilliers* expedition, for which he served as the lead scientist; there are shots of the ship, the tents, and, if you look closely enough, the bears. One grainy-looking photo shows someone dressed up as Santa Claus, celebrating Christmas in the darkness out on the ice. "The most fun you could ever have" was how Perovich described the expedition to me.

Perovich's particular area of expertise, in the words of his CRREL biography, is "the interaction of solar radiation with sea ice." During the *Des Groseilliers* expedition, Perovich spent most of his time monitoring conditions on the floe using a device known as a spectroradiometer. Facing toward the sun, a spectroradiometer measures incident light, and facing toward earth, it measures reflected light. By dividing the latter by the former, you get a quantity known as albedo. (The term comes from the Latin word for "whiteness.") During April and May, when conditions on the floe were relatively stable, Perovich took measurements with his spectroradiometer once a week, and during June, July, and August, when they were changing more rapidly, he took measurements every other day. The arrangement allowed him to plot exactly how the albedo varied as the snow on top of the ice turned to slush, and then the slush became puddles, and, finally, some of the puddles melted through to the water below.

An ideal white surface, which reflected all the light that shone on it, would have an albedo of one, and an ideal black surface, which absorbed all the light, would have an albedo of zero. The albedo of the earth, in aggregate, is 0.3, meaning that a little less than a third of the sunlight that strikes it is reflected back out. Anything that changes the earth's albedo changes how much energy the planet absorbs, with potentially dramatic consequences. "I like it because it deals with simple concepts, but it's important," Perovich told me.

At one point, Perovich asked me to imagine that we were looking down at the earth from a spaceship hovering above the North Pole. "It's springtime, and the ice is covered with snow, and it's really bright and white," he said. "It reflects over 80 percent of the incident sunlight. The albedo's around 0.8, 0.9. Now, let's suppose that we melt that ice away and we're left with the ocean. The albedo of the ocean is less than 0.1; it's like 0.07.

"Not only is the albedo of the snow-covered ice high; it's the highest of anything we find on earth," he went on. "And not only is the albedo of water low; it's pretty much as low as anything you can find on earth. So what you're doing is you're replacing the best reflector with the worst reflector." The more open water that's exposed, the more solar energy goes into heating the ocean. The result is a positive feedback, similar to the one between thawing permafrost and carbon releases, only more direct. This so-called ice-albedo feedback is believed to be a major reason that the Arctic is warming so rapidly.

"As we melt that ice back, we can put more heat into the system, which means we can melt the ice back even more, which means we can put more heat into it, and, you see, it just kind of builds on itself," Perovich said. "It takes a small nudge to the climate system and amplifies it into a big change."

A few dozen miles to the east of CRREL, not far from the Maine–New Hampshire border, is a small park called the Madison Boulder Natural Area. The park's major—indeed, only—attraction is a block of granite the size of a two-story house. The Madison Boulder is thirty-seven feet wide and eighty-three feet long and weighs about ten million pounds. It was plucked out of the White Mountains and deposited in its current location eleven thousand years ago, and it illustrates how relatively minor changes to the climate system can, when amplified, yield monumental results.

Geologically speaking, we are now living in a warm period after an ice age. Over the past two million years, huge ice sheets have advanced across the Northern Hemisphere and retreated again more than twenty times. (Each major advance tended, for obvious reasons, to destroy the evidence of its predecessors.) The most recent advance, called the Wisconsin, began roughly 120,000 years ago. Ice began to creep outward from centers in Scandinavia, Siberia, and the highlands near Hudson Bay, spreading gradually across what is now Europe and Canada. By the time the sheets had reached their maximum southern extent, most of New England and New York and a good part of the upper Midwest were buried under ice nearly a mile thick. The ice sheets were so heavy that they depressed the crust of the earth, pushing it down into the mantle. (In some places, the process of recovery, called isostatic rebound, is still going on.) As the ice retreated, at the start of the current interglacial—the Holocene—it deposited, among other landmarks, the terminal moraine known as Long Island.

It is now known, or at least almost universally accepted, that glacial cycles are initiated by slight, periodic variations in the earth's orbit. These orbital variations, which are caused by, among other things, the gravitational pull of the other planets, alter the distribution of sunlight at different latitudes during different seasons and occur according to a complex cycle that takes a hundred thousand years to complete. Orbital variations in themselves, however, aren't sufficient to produce the sort of massive ice sheet that picked up the Madison Boulder.

The crushing size of that ice sheet, the Laurentide, which stretched over some five million square miles, was the result of feedbacks, more or less analogous to those now being studied in the Arctic, only operating in reverse. As the ice spread, albedo increased, leading to less heat absorption and the growth of yet more ice. At the same time, for reasons that are not entirely understood, as the ice sheets advanced, CO_2 levels declined: during each of the most recent glaciations, carbon dioxide levels dropped almost precisely in sync with falling temperatures. During each warm period, when the ice retreated, CO_2 levels rose again. Researchers who have studied this history have concluded that fully half the temperature difference between cold periods and warm ones can be attributed to changes in the concentrations of greenhouse gases.

While I was at CRREL, Perovich took me to meet a colleague of his named John Weatherly. Posted on Weatherly's office door was a bumper sticker designed to be pasted—illicitly—on SUVs. It said, I'M CHANGING THE CLIMATE! ASK ME HOW! Weatherly is a climate modeler, and for the past several years, he and Perovich have been working to translate the data gathered on the *Des Groseilliers* expedition into computer algorithms to be used in climate forecasting. Weatherly told me that some climate models—worldwide, there are about fifteen major ones in operation—predict that the perennial sea-ice cover in the Arctic will disappear entirely by the year 2080. At that point, although there would continue to be seasonal ice that forms in winter, in summer the Arctic Ocean would be completely ice-free. "That's not in our lifetime," he observed. "But it is in the lifetime of our kids."

Later, back in his office, Perovich and I talked about the

long-term prospects for the Arctic. Perovich noted that the earth's climate system is so vast that it is not easily altered. "On the one hand, you think, It's the earth's climate system; it's big, it's robust. And, indeed, it has to be somewhat robust or else it would be changing all the time." On the other hand, the climate record shows that it would be a mistake to assume that change, when it comes, will come gradually. Perovich offered a comparison that he had heard from a glaciologist friend. The friend likened the climate system to a rowboat: "You can tip and then you'll just go back. You can tip it and just go back. And then you tip it and you get to the other stable state, which is upside down."

Perovich said that he also liked a regional analogy. "The way I've been thinking about it, riding my bike around here, is, You ride by all these pastures and they've got these big granite boulders in the middle of them. You've got a big boulder sitting there on this rolling hill. You can't just go by this boulder. You've got to try to push it. So you start rocking it, and you get a bunch of friends, and they start rocking it, and finally it starts moving. And then you realize, Maybe this wasn't the best idea. That's what we're doing as a society. This climate, if it starts rolling, we don't really know where it will stop."

Chapter 2

A WARMER SKY

A S A C A U S E for alarm, global warming could be said to be a 1970s idea; as pure science, however, it is much older than that. In the late 1850s, an Irish physicist named John Tyndall set out to study the absorptive properties of various gases. What he discovered led him to propose the first accurate account of how the atmosphere functions.

Tyndall, who was born in County Carlow in 1820, left school at the age of seventeen or eighteen and went to work as a surveyor for the British government. Pursuing his education at night, he subsequently became a mathematics teacher, and then, although he spoke no German, set off for Marburg to study with Robert Wilhelm Bunsen (for whom the Bunsen burner would later be named). After Tyndall received his Ph.D.—the degree was at the time just being established—he had trouble supporting himself until, in 1853, he was invited to deliver a single lecture at London's Royal Institution, then one of Britain's leading scientific centers. Based on the talk's success, Tyndall was invited to deliver another, and then another, and a few months later was elected to a professorship in natural philosophy. His lectures were enormously popular—many were collected and published—a fact that testifies both to Tyndall's considerable skills as a speaker and also to the intellectual interests of the Victorian middle class. Eventually, Tyndall went on a lucrative speaking tour of the United States, the proceeds from which he placed in a special trust to be used for the advancement of American science.

Tyndall's research varied almost impossibly widely, from optics to acoustics to glacial motion. (He was an avid mountain climber, and made frequent trips to the Alps to study the ice.) One of his most enduring interests was in the science of heat, which, in the mid-nineteenth century, was rapidly evolving. In 1859, Tyndall built the world's first ratio spectrophotometer, a device that allowed him to compare the way different gases absorb and transmit radiation. When Tyndall tested the most common gases in the air—nitrogen and oxygen—he found they were transparent to both visible and infrared radiation. (The latter of these he called "ultra-red" radiation.) Other gases, like carbon dioxide, methane, and water vapor, however, were not. CO_2 and water vapor were transparent in the visible part of the spectrum, but partly opaque in the infrared. Tyndall was quick to appreciate the implications of his discovery: the selectively transparent gases, he declared, were largely responsible for determining the planet's climate. He likened their impact to that of a dam built across a river: just as a dam "causes a local deepening of the stream, so our atmosphere, thrown as a barrier across the terrestrial rays, produces a local heightening of the temperature at the earth's surface."

The phenomenon that Tyndall identified is now referred to as the "natural greenhouse effect." It is not remotely controversial; indeed, it's recognized as an essential condition of life on the planet. To understand how it works, it helps to imagine the world without it. In that situation, the earth would be constantly receiving energy from the sun and, at the same time, constantly radiating energy back out to space. All hot bodies radiate, and the amount that they radiate is a function of their temperature. (The exact relationship is expressed by a formula known as

The world's first ratio spectrophotometer, built by John Tyndall, was used to measure the absorptive properties of gases. Credit: Philosophical Transactions, vol. 151 (1861).

the Stefan-Boltzmann law, which states that the radiation emitted by an object is proportional to its absolute temperature raised to the fourth power: $P/A = \sigma T^4$*.) In order for the earth to be in equilibrium, the quantity of energy it radiates out into space must equal the quantity of radiation it is receiving. When, for whatever reason, equilibrium is disturbed, the planet will either warm up or cool down until its temperature is once again sufficient to make the two energy streams balance out.

If there were no greenhouse gases in the atmosphere, energy radiating from the surface of the earth would flow away unimpeded. In that case, it would be comparatively easy to calculate how warm the planet would have to be to throw back into space the same amount of energy it receives from the sun. (This amount varies widely by location and time of year; averaged out over all latitudes and all seasons it comes to some 235 watts per square meter, or roughly the power of four household lightbulbs.) The result of this calculation turns out to be a frigid zero degrees. To use Tyndall's Victorian language, if the heat-trapping gases were removed from the earth's atmosphere, "the warmth of our fields and gardens would pour itself unrequited into space, and the sun would rise upon an island held fast in the iron grip of frost."

Greenhouse gases alter the situation because of their selectively absorptive properties. They allow the sun's radiation, which arrives mostly in the form of visible light,

---

* P stands for power, in watts; A for area, in square meters; T for temperature, in degrees Kelvin. σ is the Stefan-Boltzmann constant, 5.67×10^{-8} W/m^2K^4.

to pass freely. But the earth's radiation, which is emitted in the infrared part of the spectrum, is partially blocked. Greenhouse gases absorb infrared radiation and then re-emit it—some out toward space and some back toward earth. This process of absorption and re-emission has the effect of limiting the outward flow of energy; as a result, the earth's surface and its lower atmosphere need to be that much warmer before the planet can radiate out the necessary 235 watts per square meter. The presence of greenhouse gases largely accounts for the fact that the average global temperature, instead of zero, is actually a far more comfortable fifty-seven degrees.

Tyndall suffered from insomnia, which grew worse as he grew older, and in 1893 he died from an overdose of chloral hydrate—an early sleeping drug—that had been administered by his wife. ("My poor darling, you have killed your John," he is reported to have told her shortly before expiring.) Right around the time of his poisoning, the Swedish chemist Svante Arrhenius took up where he had left off.

Arrhenius would eventually come to be regarded as one of the giants of nineteenth-century science, but his career, like Tyndall's, began inauspiciously. In 1884, when Arrhenius was a student at the University of Uppsala, he wrote a doctoral dissertation on the behavior of electrolytes. (In 1903, he would be awarded the Nobel Prize for this work, now known as the theory of electrolytic dissociation.) The university's examining committee was so unimpressed that it awarded the dissertation a fourth-class mark: *non sine laude*. Arrhenius spent the next several years bouncing from one foreign post to another before finally being offered a teaching position back home in Sweden. He would not be elected to the Swedish Academy of Sciences until shortly before winning the Nobel Prize, and even then his election faced strong opposition.

Why, exactly, Arrhenius became curious about the effects of CO_2 on global temperatures is unclear; mainly he seems to have been interested in determining whether falling carbon dioxide levels could have caused the ice ages. (Some biographers have noted, although it's hard to find any real connection, that his work on the subject coincided with his separation from his wife—earlier his student—who had taken their only son with her.) Tyndall had recognized the influence of greenhouse gas levels on the climate, and indeed had even pro-posed—presciently, but not entirely correctly—that var-iations in these levels would have been capable of producing "all the mutations of climate which the researches of geologists reveal." But Tyndall never went beyond such qualitative speculations. Arrhenius decided to actually calculate how the earth's temperature would be affected by changing CO_2 levels. He would later describe this task as one of the most tedious of his life. He began working on it on Christmas Eve 1894, and although he routinely toiled for fourteen hours a day— "I have not worked this hard since I was cramming for my B.A.," he wrote to a friend—he was not finished for nearly a year. Finally, in December 1895, he was ready to present his conclusions to the Swedish Academy.

By today's standards, Arrhenius's work seems primitive. All of his calculations were performed using pen and paper. He was missing crucial pieces of information about spectral absorption, and he ignored several potentially important feedbacks. These deficiencies, however, seem more or less to have canceled each other out. Arrhenius asked what would happen to the earth's climate if CO_2 levels were halved and also if they were doubled. In the case of doubling, he determined that average global temperatures would rise between nine and eleven degrees, a result that approximates the estimates of the most sophisticated climate models in operation today.

Arrhenius was also responsible for a key conceptual breakthrough. All over Europe, factories and railroads and power stations were burning coal and belching out smoke. Arrhenius recognized that industrialization and climate change were intimately related, and that the consumption of fossil fuels must, over time, lead to warming. He was not, however, terribly concerned about this. Arrhenius thought that the buildup of carbon dioxide in the air would be extremely slow—at one point, he estimated that it would take three thousand years of coal burning to double atmospheric levels—mostly because he believed the oceans would act as a vast sponge, soaking up extra CO_2. Perhaps owing to the age he lived in, or perhaps just because he was Scandinavian, he anticipated that the results would, on the whole, be salubrious. Addressing the Swedish Academy,

Arrhenius declared that rising levels of carbon dioxide, which at the time was referred to as "carbonic acid," would allow future generations "to live under a warmer sky." Later, he elaborated on this notion in one of his numerous works of popular science, *Worlds in the Making*:

> By the influence of the increasing percentage of carbonic acid in the atmosphere we may hope to enjoy ages with more equable and better climates, especially as regards the colder regions of the earth, ages when the earth will bring forth much more abundant crops than at present for the benefit of rapidly propagating mankind.

After Arrhenius's death, in 1927, interest in climate change dropped off. Most scientists continued to believe that if carbon dioxide levels were rising at all, they were rising very slowly. Then, in the mid-1950s, for no particularly good reason, a young chemist named Charles David Keeling decided to work out a new and more precise way of measuring atmospheric CO_2. (Later he would explain his decision by saying he was "having fun" trying to assemble the necessary equipment.) In 1958, Keeling convinced the U.S. Weather Bureau to start using his technique to monitor CO_2 at its new observatory, eleven thousand feet above sea level, on the flank of Mauna Loa, on the island of Hawaii. These same CO_2 measurements have been taken at Mauna Loa nearly continuously ever since. The results, known as the "Keeling Curve," may well be the most widely reprinted set of natural science data ever collected.

Presented in the form of a graph, the Keeling Curve looks like the edge of a saw that is being held at a tilt. Each tooth on the saw corresponds to a single year. CO_2 levels fall to a minimum in the summer, when the trees of the Northern Hemisphere are taking up carbon dioxide for photosynthesis, and rise to a maximum in the winter, when these trees go dormant. (In the Southern Hemisphere, there are fewer forests.) The tilt, meanwhile, corresponds to the rising annual mean.

The first full year that CO_2 levels were recorded at Mauna Loa—1959—that mean stood at 316 parts per million. By the following year, it had reached 317 parts per million, prompting Keeling to observe that the reigning assumption about CO_2 absorption by the oceans was

The Keeling Curve shows that CO_2 levels have been rising steadily since the 1950s. Credit: Scripps Institution of Oceanography.

probably wrong. By 1970, the level had reached 325 parts per million, and by 1990, it was up to 354 parts per million. In the summer of 2005, the CO_2 level stood at 378 parts per million, and by now, it has almost certainly risen to 380 parts per million. At this rate, it will reach 500 parts per million—nearly double preindustrial levels—by the middle of this century, which is to say, roughly two thousand eight hundred and fifty years ahead of Arrhenius's prediction.

Chapter 3

UNDER THE GLACIER

S WISS CAMP IS a research station that was set up in
1990 on a platform drilled into the Greenland ice sheet.
Ice flows like water, only more slowly, and, as a result, the
camp is always in motion: in fifteen years, it has migrated
by more than a mile, generally in a westerly direction.
Every summer, the whole place gets flooded, and every
winter, its contents solidify. The cumulative effect of all
this is that almost nothing at Swiss Camp functions anymore the way it was supposed to. To get into it, you have
to clamber up a snowdrift and descend through a trapdoor
in the roof, as if entering a ship's hold or a space module.
The living quarters are no longer habitable, so now everyone at the camp sleeps outside, in tents. (The one assigned
to me was, I was told, the same sort used by Robert Scott
on his ill-fated expedition to the South Pole.) By the time I
arrived at the camp, in late May, someone had jackham-
mered out the center of the workspace, which was
equipped with some battered conference tables. Under
the tables, where, under normal circumstances, you would
stick your legs, there were still three-foot-high blocks of

ice. Inside of the blocks, I could dimly make out a tangle of wires, a bulging plastic bag, and an old dustpan.

Konrad Steffen, a professor of geography at the University of Colorado, is the director of Swiss Camp. A native of Zurich, Steffen speaks English in the lilting cadences of *Schweizerdeutsch*. He is tall and lanky, with pale blue eyes, a graying beard, and the unflappable manner of a cowboy in a western. Steffen fell in love with the Arctic when, as a graduate student in 1975, he spent a summer on Axel Heiberg Island, four hundred miles northwest of the north magnetic pole. A few years later, for his doctoral dissertation, he lived for two winters on the sea ice off Baffin Island. (Steffen told me that for his honeymoon he had wanted to take his wife to Spitsbergen, an island five hundred miles north of Norway, but she demurred, and they ended up driving across the Sahara instead.)

When Steffen planned Swiss Camp—he built much of the place himself—it was not with global warming in mind. Rather, he was interested in following meteorological conditions on what is known as the ice sheet's "equilibrium line." Along this line, winter snow and summer melt are supposed to be precisely in balance. But in recent years, "equilibrium" has become an increasingly elusive quality. During the summer of 2002, for example, melt occurred in areas where liquid water had not been seen for hundreds, perhaps thousands, of years. The following winter, there was an unusually low snowfall, and in the summer of 2003, the melt was so great that, around Swiss Camp, five feet of ice were lost.

When I arrived at the camp, the 2004 melt season was already under way. This, to Steffen, was a matter of both intense scientific interest and serious practical concern. A few days earlier, one of his graduate students, Russell Huff, and a postdoc, Nicolas Cullen, had driven out on snowmobiles to service some weather stations closer to the coast. The snow there was warming so fast that they had had to work until five in the morning, and then take a long detour back, to avoid getting caught in the quickly forming rivers. Steffen wanted to complete everything that needed to be done ahead of schedule, in case everyone had to pack up and leave early. My first day at Swiss Camp he spent fixing an antenna that had fallen over in the previous year's melt. It was bristling with equipment, like a high-tech Christmas tree. Even on a relatively mild day on the ice sheet, which this was, it never gets more than a few degrees above freezing, and I was walking around in a huge parka, two pairs of pants plus long underwear, and two pairs of gloves. Steffen, meanwhile, was tinkering with the antenna with his bare hands. He had spent the last fourteen summers at Swiss Camp, and I asked him what he had learned during that time. He answered with another question.

"Are we disintegrating part of the Greenland ice sheet over the longer term?" he asked. He was sorting through a tangle of wires that to me all looked the same but must have had some sort of distinguishing characteristics. "What the regional models tell us is that we will get more melt at the coast. It will continue to melt. But warmer air can hold more water vapor, and at the top of the ice sheet you'll get more precipitation. So we'll add more snow there. We'll get an imbalance of having more accumulation at the top, and more melt at the bottom. The key question now is: What is the dominant one, the more melt or the increase?"

Greenland, the world's largest island, is nearly four times the size of France—840,000 square miles—and, except for its southern tip, lies entirely above the Arctic Circle. The first Europeans to make a stab at settling it were the Norse, under the leadership of Erik the Red, who, perhaps deliberately, gave the island its misleading name. In the year 985, he arrived with twenty-five ships and nearly seven hundred followers. (Erik had left Norway when his father was exiled for killing a man, and then was himself exiled from Iceland for killing several more.) The Norse established two settlements: the Eastern Settlement, which was actually in the south, and the Western Settlement, which was to the north. For roughly four hundred years, they managed to scrape by, hunting, raising livestock, and making occasional logging expeditions to the coast of Canada. But then something went wrong. The last written record of them is an Icelandic affidavit regarding the marriage of Thorstein Ólafsson and Sigridur Björnsdóttir, which took place in the Eastern Settlement on the "Second Sunday after the Mass of the Cross," in the autumn of 1408.

These days the island has just over fifty-six thousand inhabitants, most of them Inuit, and almost a quarter live in the capital, Nuuk, about four hundred miles up the western coast. Since the late 1970s, Greenland has enjoyed a measure of home rule, but the Danes, who consider the island a province, still spend more than three hundred million dollars a year to support it. The result is a thin and not entirely convincing first-world veneer. Greenland has almost no agriculture, or industry, or, for that matter, roads. Following Inuit tradition, private ownership of land is not allowed, although it is possible to buy a house, an expensive proposition in a place where even the sewage pipes have to be insulated.

More than 80 percent of Greenland is covered by ice. Locked into this enormous glacier is 8 percent of the world's fresh water supply. Except for researchers like those at Swiss Camp, no one lives on the ice, or even ventures out onto it very often. (The edges are riddled with crevasses large enough to swallow a dogsled or, should the occasion arise, a five-ton truck.)

Like all glaciers, the Greenland ice sheet is made up entirely of accumulated snow. The most recent layers are thick and airy, while the older layers are thin and dense, which means that to drill through the ice is to descend backward in time, at first gradually, and then much more rapidly. A hundred and thirty-eight feet down, there is snow that fell during the time of the American Civil War; 2,500 feet down, snow from the time of the Peloponnesian Wars, and, 5,350 feet down, snow from the days when the cave painters of Lascaux were slaughtering bison. At the very bottom, 10,000 feet down, there is snow that fell on central Greenland before the start of the last ice age, more than a hundred thousand years ago.

As the snow is compressed, its crystal structure changes to ice. (Two thousand feet down, there is so much pressure on the ice that a sample drawn to the surface will, if mishandled, fracture, and in some cases even explode.) But in most other respects, the snow remains unchanged, a relic of the climate that first formed it. In the Greenland ice, there is nuclear fallout from early atomic tests, volcanic ash from Krakatau, lead pollution from ancient Roman smelters, and dust blown in from Mongolia on ice age winds. Every layer also contains tiny bubbles of trapped air, each of them a sample of a past atmosphere.

Much of what is known about the earth's climate over the last hundred thousand years comes from ice cores drilled in central Greenland, along a line known as the ice divide. Owing to differences between summer and winter snow, each layer in a Greenland core can be individually dated, much like the rings of a tree. Then, by analyzing the isotopic composition of the ice, it is possible to determine how cold it was at the time each layer was formed. Over the last decade, three Greenland cores have been drilled to a depth of nearly two miles, and these cores have prompted a wholesale rethinking of how the climate operates. Where once the system was thought to change, as it were, only glacially, now it is known to be capable of sudden and unpredictable reversals. One such reversal, called the Younger Dryas, after a small Arctic plant—*Dryas octopetala*—that suddenly reappeared in Scandinavia, took place roughly 12,800 years ago.

At that point, the earth, which had been warming rapidly, was plunged back into ice age conditions. It remained frigid for twelve centuries and then warmed again, even more abruptly. In Greenland, average annual temperatures shot up by nearly twenty degrees in a single decade.

As a continuous temperature record, the Greenland ice cores stop providing reliable information right around the start of the last glaciation. Climate records pieced together from other sources indicate that the previous interglacial,

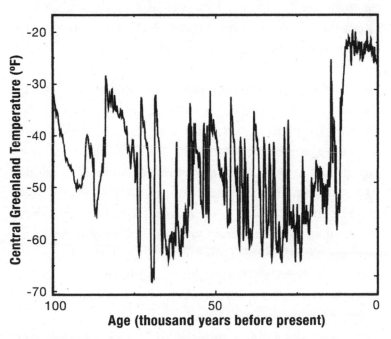

The Greenland record reveals that temperatures have often swung wildly. Credit: The Two-Mile Time Machine, *Princeton University Press, after K. Cuffey and G. Clow,* Journal of Geophysical Research, *vol. 102 (1997).*

which is known as the Eemian, was somewhat warmer than the present one, the Holocene. They also show that sea levels during that time were at least fifteen feet higher than they are today. One theory attributes this to a collapse of the West Antarctic ice sheet. A second holds that meltwater from Greenland was responsible. (When sea ice melts, it does not affect sea level, because the ice, which was floating, was already displacing an equivalent volume of water.) All told, the Greenland ice sheet holds enough water to raise sea levels worldwide by twenty-three feet. Scientists at NASA have calculated that throughout the 1990s the ice sheet, despite some thickening at the center, was shrinking by twelve cubic miles per year.

Jay Zwally is a NASA scientist who works on a satellite project known as the Ice Cloud and Land Elevation Satellite (ICESat). He is short and stocky, with a round face and a mischievous grin. Zwally is a friend of Steffen's and about ten years ago, he got the idea of installing global-positioning-system receivers around Swiss Camp to study changes in the ice sheet's elevation. He happened to be at the camp at the same time I was, and the second day of my visit we all got onto snowmobiles and headed out to a location known as JAR 1 (for Jakobshavn Ablation Region) to reinstall a GPS receiver. The trip was about ten miles. Midway through it, Zwally told me that he had once seen spy-satellite photos of the region we were crossing, and that they had shown that underneath the snow it was full of crevasses. Later, when I asked Steffen about this, he told me that he had had the whole area surveyed with bottom-seeking radar, and no crevasses of any note had been found. I was never sure which one of them to believe.

Reinstalling Zwally's GPS receiver entailed putting up a series of poles, a process that, in turn, required drilling holes thirty feet down into the ice. The drilling was done not mechanically but thermally, using a steam drill that consisted of a propane burner, a steel tank, and a long rubber hose. Everyone—Steffen, Zwally, the graduate students, me—took a turn. This meant holding on to the hose while it melted its way down, an activity reminiscent of ice fishing. Seventy-five years ago, not far from JAR 1, Alfred Wegener, the German scientist who proposed the theory of continental drift, died while on a meteorological expedition. He was buried in the ice sheet, and there is a running joke at Swiss Camp about stumbling onto his body. "It's Wegener!" one of the graduate students exclaimed, as the drill worked its way downward. The first hole was finished relatively quickly, at which point everyone decided—prematurely, as it turned out—that it was time for a midday break. Unless a hole stays filled with water, it starts to close up again, and can't be used. Apparently, there were fissures in the ice, because water kept draining out of the next few holes that were tried. The original plan had been for three holes, but, some six hours later, only two had been drilled, and it was decided that this would have to suffice.

Although Zwally had set out to look for changes in the

ice sheet's elevation, what he ended up discovering was even more significant. His GPS data showed that as the ice sheet melted, it didn't so much sink as start to accelerate. Thus, in the summer of 1996, the ice around Swiss Camp moved at a rate of thirteen inches per day, but, in 2001, it had sped up to twenty inches per day. The reason for this acceleration, it is believed, is that meltwater from the surface makes its way down to the bedrock below, where it acts as a lubricant. (In the process, it enlarges cracks and forms huge ice tunnels, known as "moulins.") Zwally's measurements also showed that, in the summer, the ice sheet rises by about six inches, indicating that it is floating on a cushion of water.

At the end of the last glaciation, the ice sheets that covered much of the Northern Hemisphere disappeared in a matter of a few thousand years—a surprisingly short time, considering how long it had taken them to build up. At one point, about fourteen thousand years ago, they were melting so fast that sea levels were rising at the rate of more than a foot a decade. Just how this happened is not entirely understood, but the acceleration of the Greenland ice sheet suggests yet another feedback mechanism: once an ice sheet begins to melt, it starts to flow faster, which means it also thins out faster, encouraging further melt. Not far from Swiss Camp is the huge river of ice known as the Jakobshavn Isbrae. In 1992, the Jakobshavn Isbrae flowed at a rate of 3.5 miles per year; by 2003, its velocity had increased to 7.8 miles per year. (Similar findings were announced recently by scientists measuring the flow of ice streams on the Antarctic Peninsula.) On the basis of Zwally's findings, James Hansen, the NASA official who directed one of the initial 1970s studies on the effects of carbon dioxide, has argued that if greenhouse gas emissions are not controlled, the total disintegration of the Greenland ice sheet could be set in motion in a matter of decades. Although the process could take centuries to fully play out, once begun it would become self-reinforcing, and hence virtually impossible to stop. In an article published in the journal *Climatic Change* in February 2005, Hansen, who is now the head of the Goddard Institute for Space Studies, wrote that he hoped he was wrong about the ice sheet, but added, "I doubt it."

As it happened, I was at Swiss Camp just as the global-warming disaster movie *The Day After Tomorrow* was opening in theaters. One night, Steffen's wife called on the camp's satellite phone to say that she had just taken the couple's two teenage children to see it. Everyone had enjoyed the film, she reported, especially because of the family connection.

The fantastic conceit of *The Day After Tomorrow* is that global warming produces global freezing. At the start of the film, a chunk of Antarctic ice the size of Rhode Island suddenly melts. (Something very similar to this actually happened in March 2002, when the Larsen B ice shelf collapsed.) Most of what follows—an instant ice age, cyclonic winds that descend from the upper atmosphere—is impossible as science but not as metaphor.

The record preserved in the Greenland ice sheet shows that our own relatively static experience of climate is actually what is exceptional. During the last glaciation, even as much of the world was frozen solid, average temperatures in Greenland frequently shot up, or down, by ten degrees, as in the Younger Dryas. Nobody knows what caused the sudden climate shifts of the past; however, many climatologists suspect that they had something to do with changes in ocean-current patterns that are known as the "thermohaline circulation."

"When you freeze sea ice, the salt is pushed out of the pores, so that the salty water actually drains," Steffen explained to me one day when we were standing out on the ice, not far from camp, trying to talk above the howl of the wind. "And salty water's actually heavier, so it starts to sink." Meanwhile, owing both to evaporation and cooling, water from the tropics becomes denser as it drifts toward the Arctic; near Greenland a tremendous volume of seawater is constantly sinking toward the ocean floor. As a result of this process, still more warm water is drawn from the tropics toward the poles, setting up what is often referred to as a "conveyor belt" that moves vast amounts of heat around the globe.

"This is the energy engine for the world climate," Steffen went on. "And it has one source: the water that sinks down. And if you just turn the knob here a little bit"—he made a motion of turning the water on in a bathtub—"we can expect significant temperature changes based on the redistribution of energy." One way to turn the knob is to heat the oceans, which is already happening. Another is to pour more freshwater into the polar seas. This is also occurring. Not only is runoff from coastal Greenland increasing; the volume of river discharge into the Arctic Ocean has been rising. Oceanographers monitoring the North Atlantic have documented that in recent decades its waters have become significantly less salty. A total shutdown of the thermohaline circulation is considered extremely unlikely in the coming century. But, if the Greenland ice sheet were to start to disintegrate, the possibility of such a shutdown could not be ruled out. Wallace Broecker, a professor of geochemistry at Columbia University's Lamont-Doherty Earth Observatory, has labeled the thermohaline circulation the "Achilles' heel of the climate system." Were it to halt, places like Britain, whose climate is heavily influenced by the Gulf Stream, could become much colder, even as the planet as a whole continued to warm up.

For the whole time I was at Swiss Camp, it was "polar day," and so the sun never set. Dinner was generally served at ten or eleven P.M., and afterward everyone sat around a makeshift table in the kitchen, talking and drinking coffee. (Because it weighs a lot and is not—strictly speaking—necessary, alcohol was in short supply.) One night, I asked Steffen what he thought conditions at Swiss Camp would be like in the same season a decade hence. "In ten years, the signal should be much more distinct, because we will have added another ten years of greenhouse warming," he said.

Zwally interjected, "I predict that ten years from now we won't be coming this time of year. We won't be able to come this late. To put it nicely, we are heading into deep doo-doo."

Either by disposition or by training, Steffen was reluctant to make specific predictions, whether about Greenland or, more generally, about the Arctic. Often, he prefaced his remarks by noting that there could be a change in atmospheric-circulation patterns that would dampen the rate of temperature increase or even—temporarily, at least—reverse it entirely. But he was emphatic that "climate change is a real thing."

"It's not something dramatic now—that's why people don't really react," he told me. "But if you can convey the message that it will be dramatic for our children and our children's children—the risk is too big not to care." The time, he added, "is already five past midnight."

On the last night that I spent at Swiss Camp, Steffen took the data he had downloaded off his weather station and ran them through various programs on his laptop to produce the mean temperature at the camp for the previous year. It was, it turned out, the highest of any year since the camp was built. When Steffen announced this to the group around the kitchen table, no one seemed the slightest bit surprised.

That night, dinner was unusually late. On the return trip of another pole-drilling expedition, one of the snowmobiles had caught on fire, and had had to be towed back to camp. When I finally went out to my tent to go to bed, I found that the snow underneath it had started to melt, and there was a large puddle in the middle of the floor. I went back to the kitchen to get some paper towels and tried to mop it up. But the puddle was too big, and eventually I gave up.

No nation takes a keener interest in climate change, at least on a per-capita basis, than Iceland. More than 10 percent of the country is covered by glaciers, the largest of which, Vatnajökull, stretches over thirty-two hundred square miles. During the so-called Little Ice Age, which began in Europe some five hundred years ago and ended some three hundred and fifty years later, the advance of the glaciers caused widespread misery. Contemporary records tell of farms being buried under the ice—"Frost and cold torment people," a pastor in eastern Iceland named Olafur Einarsson wrote—and in particularly severe years, shipping, too, seems to have ceased, because the island remained icebound even in summer. In the mid-eighteenth century, it has been estimated, nearly a third of the country's population died of starvation or associated cold-related ills. For Icelanders, many of whom can trace their genealogy back a thousand years, this is considered to be almost recent history.

Oddur Sigurdsson heads up a group called the Icelandic Glaciological Society. On a dark and dreary autumn afternoon, I went to visit him in his office, at the headquarters of Iceland's National Energy Authority, in Reykjavík. Little towheaded children kept wandering in to peer under his desk, and then wandering out again, giggling. Sigurdsson explained that Reykjavík's public school teachers were on strike, and his colleagues had had to bring their children to work with them.

The Icelandic Glaciological Society is composed entirely of volunteers. Every fall, after the summer-melt season has ended, they survey the size of the country's three-hundred-odd glaciers and then file reports, which Sigurdsson collects in brightly colored binders. In the organization's early years—it was founded in 1930—the volunteers were mostly farmers; they took measurements by building cairns and pacing off the distance to the glacier's edge. These days, members come from all walks of life—one is a retired plastic surgeon—and they take more exacting surveys, using tape measures and iron poles. Some glaciers have been in the same family, so to speak, for generations. Sigurdsson became head of the society in 1987, at which point one volunteer told him that he thought he would like to relinquish his post.

"He was about ninety when I realized how old he was," Sigurdsson recalled. "His father had done this at that place before and then his nephew took over for him." Another volunteer has been monitoring his glacier, a section of Vatnajökull, since 1948. "He's eighty," Sigurdsson said. "And if I have some questions that go beyond his age, I just go and ask his mother. She's a hundred and seven."

In contrast to glaciers in North America, which have been shrinking steadily since the 1960s, Iceland's glaciers grew through the 1970s and '80s. Then, in the mid-1990s, they, too, began to contract. Sigurdsson pulled out a notebook of glaciological reports, filled out on yellow forms, and turned to the section on a glacier called Sólheimajökull, a tongue-shaped spit of ice that sticks out from a much larger glacier known as Mýrdalsjökull. In 1996, Sólheimajökull crept back by 10 feet. In 1997, it receded by another 33 feet, and in 1998 by 98 feet. Every year since then, it has retreated even more. In 2003, it shrank by 302 feet, and in 2004, by 285 feet. All told, Sólheimajökull—the name means "sun-home glacier" and refers to a nearby farm—is now 1,100 feet shorter than it was just a decade ago. Sigurdsson pulled out another notebook, which was filled with slides. He picked out some recent ones of Sólheimajökull. The glacier ended in a wide river. An enormous rock, which Sólheimajökull had deposited when it began its retreat, stuck out from the water like the hull of an abandoned ship.

"You can tell by this glacier what the climate is doing," Sigurdsson said. "It is more sensitive than the most sensitive meteorological measurement." He introduced me to a colleague of his, Kristjana Eythórsdóttir, who, as it turned out, was the granddaughter of the founder of the Icelandic Glaciological Society. Eythórsdóttir keeps tabs on a glacier named Leidarjökull, which is a four-hour trek from the nearest road. I asked her how it was doing. "Oh, it's getting smaller and smaller, just like all the others," she said. Sigurdsson told me that climate models predicted that by the end of the next century Iceland would be virtually ice-free. "We will have small ice caps on the highest mountains, but the mass of the glaciers will have gone," he said. It is believed that there have been glaciers on Iceland for at least the last two million years. "Probably longer," Sigurdsson said.

In October 2000, in a middle school in Barrow, Alaska, officials from the eight Arctic nations—the United States, Russia, Canada, Denmark, Norway, Sweden, Finland, and Iceland—met to talk about global warming. The group announced plans for a three-part, two-million-dollar study of climate change in the region. In November 2004, the first two parts of the study—a massive technical document and a hundred-and-forty-page summary—were presented at a symposium in Reykjavík.

The day after I went to talk to Sigurdsson, I attended the symposium's plenary session. In addition to nearly three hundred scientists, it drew a sizable contingent of native Arctic residents—reindeer herders, subsistence hunters, and representatives of groups like the Inuvialuit Game Council. In among the shirts and ties, I spotted two men dressed in the brightly colored tunics of the Sami and several others wearing sealskin vests. As the session went on, the subject kept changing—from hydrology and bio-diversity to fisheries and on to forests. The message, however, stayed the same. Almost wherever you looked, conditions in the Arctic were changing, and at a rate that surprised even those who had expected to find clear signs of warming. Robert Corell, an American oceanographer and former assistant director at the National Science

Foundation, coordinated the study. In his opening remarks, he ran through its findings—shrinking sea ice,
receding glaciers, thawing permafrost—and summed them
up as follows: "The Arctic climate is warming rapidly now,
with an emphasis on *now*." Particularly alarming, Corell
said, were the most recent data from Greenland, which
showed the ice sheet melting much faster "than we
thought possible even a decade ago."

Global warming is routinely described as a matter of
scientific debate—a theory whose validity has yet to be
demonstrated. The symposium's opening session lasted
for more than nine hours. During that time, many speakers stressed the uncertainties that remain about global
warming and its effects—on the thermohaline circulation,
on the distribution of vegetation, on the survival of cold-
loving species, on the frequency of forest fires. But this
sort of questioning, which is so basic to scientific discourse, never extended to the relationship between
carbon dioxide and rising temperatures. The study's executive summary stated, unequivocally, that human
beings had become the "dominant factor" influencing
the climate. During an afternoon coffee break, I caught
up with Corell.

"Let's say that there's three hundred people in this
room," he told me. "I don't think you'll find five who
would say that global warming is just a natural process."
(While I was at the conference, I spoke to more than
twenty scientists, and I couldn't find one who described it
that way.)

The third part of the Arctic-climate study, which was still unfinished at the time of the symposium, was the so-called policy document. This was supposed to outline practical steps to be taken in response to the scientific findings, including—presumably—reducing greenhouse gas emissions. The policy document remained unfinished because American negotiators had rejected much of the language proposed by the seven other Arctic nations. (A few weeks later, the United States agreed to a vaguely worded statement calling for "effective"— but not obligatory—actions to combat the problem.) This recalcitrance left those Americans who had traveled to Reykjavík in an awkward position. A few tried— halfheartedly—to defend the Bush administration's stand to me; most, including many government employees, were critical of it. At one point, Corell observed that the loss of sea ice since the late 1970s was equal to "the size of Texas and Arizona combined. That analogy was made for obvious reasons."

That evening, at the hotel bar, I talked to an Inuit hunter named John Keogak, who lives on Banks Island, in Canada's Northwest Territories, some five hundred miles north of the Arctic Circle. He told me that he and his fellow hunters had started to notice that the climate was changing in the mid-eighties. Then, a few years ago, for the first time, people began to see robins, a bird for which the Inuit in his region have no word.

"We just thought, Oh, gee, it's warming up a little bit," he recalled. "It was good at the start—warmer winters, you know—but now everything is going so fast. The things that we saw coming in the early nineties, they've just multiplied.

"Of the people involved in global warming, I think we're on top of the list of who would be most affected," Keogak went on. "Our way of life, our traditions, maybe our families. Our children may not have a future. I mean, all young people, put it that way. It's just not happening in the Arctic. It's going to happen all over the world. The whole world is going too fast."

The symposium in Reykjavík lasted for four days. One morning, when the presentations on the agenda included "Char as a Model for Assessing Climate Change Impacts on Arctic Fishery Resources," I decided to rent a car and take a drive. In recent years, Reykjavík has been expanding almost on a daily basis, and the old port city is now surrounded by rings of identical, European-looking suburbs. Ten minutes from the car-rental place, these began to give out, and I found myself in a desolate landscape in which there were no trees or bushes or really even soil. The ground—fields of lava from some defunct, or perhaps just dormant, volcanoes—resembled macadam that had recently been bulldozed. I stopped to get a cup of coffee in the town of Hveragerdi, where roses are raised in greenhouses heated with steam that pours directly out of the earth. Farther on, I crossed into farm country; the landscape was still treeless, but now there was grass, and sheep eating it. Finally, I reached the sign for Sólheimajökull, the glacier whose retreat Oddur Sigurdsson had described to me. I turned off onto a dirt road. It ran alongside a brown river, between two crazily shaped ridges. After a few miles, the road ended, and the only option was to continue on foot.

By the time I got to the lookout over Sólheimajökull, it was raining. In the gloomy light, the glacier appeared less sublime than merely forlorn. Much of it was gray—covered in a film of dark grit. In its retreat, it had left behind ridged piles of silt. These were jet-black and barren—not even the tough local grasses had had a chance to take root on them. I looked around for the enormous boulder I had seen in the photos in Sigurdsson's office. It was such a long way from the edge of the glacier that for a moment I wondered if perhaps it had been carried along by the current. A raw wind came up, and I started to head down. Then I thought about what Sigurdsson had told me. If I returned in another decade, the glacier would probably no longer even be visible from the ridge where I was standing. So I climbed back up to take a second look.

Chapter 4

THE BUTTERFLY AND THE TOAD

POLYGONIA C-ALBUM, generally known as the Comma butterfly, spends most of its life pretending to be something that it is not. In its larval, or caterpillar, stage, it has a chalky stripe down its back, which makes it look, uncannily, like a bird dropping. As an adult, with wings folded, it is practically indistinguishable from a dead leaf. The Comma gets its name from a tiny white mark on its underside shaped like the letter "C." Even this is thought to be part of its camouflage—an ersatz tear of the sort leaves get when they are particularly old and tatty.

The Comma is a European butterfly—its American cousins are the Eastern Comma and the Question Mark—and it can be found in France, where it is known as *le Robert-le-Diable*; Germany, where it is called *der C-Falter*, and the Netherlands, where it is *Gehakkelde Aurelia*. The Comma reaches the northern edge of distribution in England. This is unremarkable—many European butterflies come to the end of their range in Britain—but from a scientific standpoint fortunate.

The English have been watching and collecting butterflies for centuries—some of the specimens in the British Natural History Museum date back to the 1700s—and in the Victorian era, passion for the hobby was such that every city, and many a small town, supported its own entomological society. In the 1970s, Britain's Biological Records Centre decided to marshal this enthusiasm for a project called the Lepidoptera Distribution Maps Scheme, whose aim was to chart precisely where each of the country's fifty-nine native species could—and could not—be found. More than two thousand amateur lepidopterists participated, and in 1984, the results were collated into a hundred-and-fifty-eight-page atlas. Every species got its own map with different colored dots showing the number of times it had been sighted in any given ten square kilometers. In the map for *Polygonia c-album*, the Comma's range was shown to extend from the south coast of England northward to Liverpool in the west and Norfolk in the east. Almost immediately, this map became out of date; in the years that followed, hobbyists kept finding the Comma in new areas. By the late 1990s, the butterfly was frequently being sighted in the north of England, near Durham. By now it is well established in southern Scotland, and has been sighted as far north as the Scottish Highlands. The rate of the Comma's expansion— some fifty miles per decade—was described by the authors of the most recent butterfly atlas as "remarkable."

Chris Thomas is a biologist at the University of York who studies lepidoptera. He is tall and rangy, with an Ethan Hawke–style goatee and an amiably harried manner.

The day I met him, he had just returned from looking for butterflies in Wales, and the first thing he said to me when I got into his car was please not to mind the smell of wet socks. A few years ago, Thomas, together with his wife, their two sets of twins, an Irish wolfhound, a pony, some rabbits, a cat, and several chickens moved into an old farmhouse in the town of Wistow, in the vale of York. The University of York has an array of thermostatic chambers where Commas are raised under temperature-controlled conditions, fed carefully monitored diets, and measured on a near-constant basis, but in the spirit of British amateurism, Thomas decided to turn his own backyard into a field lab. He scattered wildflower seeds he had collected from nearby meadows and ditches, planted nearly seven hundred trees, and waited for the butterflies to show up. When I visited the place in mid-summer, the wildflowers were in bloom and the grass was so high that many of the tiny trees looked lost, like kids in search of their parents. The vale of York is almost completely flat—during the last ice age, it formed the bottom of a giant lake—and from the yard Thomas pointed out the spires of Selby Abbey, built nearly a thousand years ago, and also the cooling towers of the Drax power plant, Britain's largest, some fifteen miles away. It was cloudy, and since butterflies don't fly when it's gray, we went inside.

Butterflies, Thomas explained after putting the kettle on for tea, can be divided into two groups. First, there are the "specialists," who require specific—in some cases, unique—conditions. These include the Chalkhill Blue (*Polyommatus coridon*), a large, turquoise butterfly that feeds exclusively on horseshoe vetch, and the Purple Emperor (*Apatura iris*), which flies in the treetops of well-wooded areas in southern England. Then there are the "generalists," who are less picky. Among Britain's generalists, there are, in addition to the Comma, ten species that are widespread in the southern part of the country and reach the edge of their range somewhere in the nation's midsection. "Every single one has moved northward since 1982," Thomas told me. A few years ago, together with lepidopterists from, among other places, the United States, Sweden, France, and Estonia, Thomas conducted a survey of all the studies that had been done on generalists that reach the upper limits of their ranges in Europe. The survey looked at thirty-five species in all. Of these, the scientists found, twenty-two had shifted their range northward in recent decades, while only one had shifted south.

After a while, the sun emerged, and we went back outside. Thomas's wolfhound, Rex, a dog the size of a small horse, trailed behind us, panting heavily. Within about five minutes, Thomas had identified a Meadow Brown (*Maniola jurtina*), a Small Tortoiseshell (*Aglais urticae*), and a Green-veined White (*Pieris napi*), all species that have been flitting around Yorkshire since butterfly record-keeping began. Thomas also spotted a Gatekeeper (*Pyronia tithonus*) and a Small Skipper (*Thymelicus sylvestris*), which until recently had been confined to a region well south of where we were standing. "So far, two out of the five species of butterflies that we've seen are northward invaders," he told me. "Sometime within the last thirty years they have spread into this area." A few minutes later, he pointed out another invader sunning itself in the grass—a *Polygonia c-album*. With its wings closed, the Comma was a dull, dead-leaf brown, but with them open, it was a brilliant orange.

That life on earth changes with the climate has been assumed to be the case for a long time, indeed for very nearly as long as the climate has been known to be capable of changing. Louis Agassiz published *Études sur les glaciers*, the work in which he laid out his theory of the ice ages, in 1840. By 1859, Charles Darwin had incorporated Agassiz's theory into his own theory of evolution. Toward the end of *On the Origin of Species*, in a chapter titled "Geographical Distribution," Darwin describes the vast migrations that he supposes the advance and retreat of the glaciers must have necessitated:

> As the cold came on, and as each more southern zone became fitted for arctic beings and ill-fitted for their former more temperate inhabitants, the latter would be supplanted and arctic productions would take their places. The inhabitants of the more temperate regions would at the same time travel southward . . . As the warmth returned, the arctic forms would retreat north-ward, closely followed up in their retreat by the pro-ductions of the more temperate regions. And as the snow melted from the bases of the mountains, the arctic forms would seize on the cleared and thawed ground,

always ascending higher and higher as the warmth increased, whilst their brethren were pursuing their northern journey.

For Darwin and his contemporaries such a narrative was necessarily speculative. Much as the existence of ice ages had had to be inferred from the signs they left behind— erratics, moraines, and striated bedrock—so, too, the succession and redistribution of species on Earth could only be reconstructed from fragmentary traces: scattered bones, fossilized insect casings, ancient pollen deposits. Even as paleontologists and paleobotanists found more and more evidence of how species had responded to climate change in the past, it was taken for granted that the process was not something that could be observed in real time, an assumption that has now been proven false.

Almost anywhere you go in the world today, except perhaps for the urban areas where most of us live, it is possible to observe biological changes comparable to the northern expansion of the Comma. A recent study of common frogs living near Ithaca, New York, for example, found that four out of six species were calling—which is to say, mating—at least ten days earlier than they used to, while at the Arnold Arboretum, in Boston, the date of peak blooming for spring-flowering shrubs has advanced, on average, by eight days. In Costa Rica, birds like the keel-billed toucan (*Ramphastos sulfuratus*), once confined to the lowlands, have started to nest on mountain slopes; in the Alps, plants like purple saxifrage (*Saxifraga oppositifolia*) and Austrian draba (*Draba fladnizensis*) have been creeping up toward the summits; and in the Sierra Nevada of California, the average Edith's Checkerspot butterfly (*Euphydryas editha*) can now be found at an elevation three hundred feet higher than it was a hundred years ago. Any one of these changes could, potentially, be a response to purely local conditions—shifts, say, in regional weather patterns or in patterns of land use. The only explanation that anyone has proposed that makes sense of them all, though, is global warming.

The Bradshaw-Holzapfel Lab occupies a corner on the third floor of Pacific Hall, a peculiarly unlovely building on the campus of the University of Oregon in Eugene. At one end of the lab is a large room stacked with glassware and at the other, a pair of offices. In between are several workrooms that look, from the outside, like walk-in refrigerators. Taped to the door of one of them is a handwritten sign: "Warning—if you enter this room mosquitoes will suck your blood out through your eyes!"

William Bradshaw and Christina Holzapfel, who run the lab and share one of the offices, are evolutionary biologists. They were introduced as graduate students at the University of Michigan, and have been married for thirty-five years. Bradshaw is a tall man with thinning gray hair and a gravelly voice. His desk is covered in a mess of papers, books, and journals, and when visitors come to the lab, he likes to show them his collection of curiosities, which includes a desiccated octopus. Holzapfel is short, with blond hair and bright blue eyes. Her desk is perfectly neat.

Bradshaw and Holzapfel have shared an interest in mosquitoes for as long as they have been interested in each other. In the early years of their lab, which they set up in 1971, they raised several species, some of which, in order to reproduce, required what is delicately referred to as a "blood meal." This, in turn, demanded a live animal able to provide such a meal. For a time, this requirement was met by rats sedated with phenobarbital, but, as rules about experimenting with animals grew more stringent, Bradshaw and Holzapfel found themselves forced to decide whether it was more humane to keep sedating the same rat over and over again, or to use a new rat and let the old one wake up to find itself covered with bites. Eventually, they grew weary of such questions and decided to stick to a single species, *Wyeomyia smithii*, which needs no blood in order to reproduce. At any given moment the Bradshaw-Holzapfel Lab houses upwards of a hundred thousand *Wyeomyia smithii* in various stages of development.

Wyeomyia smithii is a small and rather ineffectual bug. ("Wimpy" is how Bradshaw characterizes it.) Its eggs are practically indistinguishable from specks of dust; its larvae appear as minuscule white worms. As an adult, it is about a quarter of an inch long and in flight looks like a tiny black blur. It is only when you examine a *Wyeomyia smithii* very closely, under a magnifying glass, that you can see that its abdomen is actually silver, and that its two hind legs are bent gracefully above its head, like a trapeze artist's.

Wyeomyia smithii completes virtually its entire life cycle—from egg to larva to pupa to adult—inside a single plant, *Sarracenia purpurea* or, as it is more commonly known, the purple pitcher plant. The purple pitcher plant, which grows in swamps and peat bogs from Florida to northern Canada, has frilly, cornucopia-shaped leaves that sprout directly out of the ground and then fill with water. In the spring, female *Wyeomyia smithii* lay their eggs one at a time, carefully depositing each in a different pitcher plant. When flies and ants and occasionally small frogs drown in the leaves of the pitcher plant—*Sarracenia purpurea* is carnivorous—their remains also provide nutrients for developing mosquito larvae. (*Sarracenia purpurea* does not digest its own food; it leaves this task to bacteria, which don't attack the mosquitoes.) When the young mature into adults, they repeat the whole process, and if conditions are favorable, the cycle can be completed four or five times in a single summer. Come fall, the adult mosquitoes die off, but the larvae live on through the winter in a state known as diapause—the insect version of hibernation.

The exact timing of diapause is critical to the survival of *Wyeomyia smithii* and also to Bradshaw and Holzapfel's research. In contrast to most insects, which rely on a variety of signals, including temperature and food availability, to regulate the onset of dormancy, *Wyeomyia smithii* depends exclusively on light cues. When the larvae perceive that day length has dropped below a certain threshold, they stop growing and molting; when they perceive that it has lengthened sufficiently, they take up again where they left off.

This light threshold, which is known as the critical photoperiod, varies from bog to bog. At the southern end of the mosquitoes' range, near the Gulf of Mexico, conditions remain favorable for breeding well into fall. A typical *Wyeomyia smithii* from Florida or Alabama will, consequently, not go dormant until day length has shrunk to about twelve and a half hours, which at that latitude corresponds to early November. At the far northern edge of the range, meanwhile, winter arrives much earlier, and an average mosquito from Manitoba will go into dormancy in late July, as soon as day length drops below sixteen and a half hours. Interpreting light cues is a genetically controlled and highly heritable trait: *Wyeomyia smithii* are programmed to respond to day length the same way their parents did, even if they find themselves living under very different conditions. (One of the walk-in-freezer-like rooms in the Bradshaw-Holzapfel Lab contains locker-size storage units, each equipped with a timer and a fluorescent bulb, where mosquito larvae can be raised under any imaginable schedule of lightness and dark.) In the mid-1970s, Bradshaw and Holzapfel demonstrated that *Wyeomyia smithii* living at different elevations also obey different light cues—high-altitude mosquitoes behave as if they were born farther north—a discovery that today might seem relatively unremarkable but at the time was sufficiently noteworthy to make the cover of *Nature*.

About five years ago, Bradshaw and Holzapfel began to

wonder about how *Wyeomyia smithii* might be affected by global warming. They knew that the species had expanded northward after the end of the last glaciation, and that at some point in the intervening millennia, the critical photoperiods of northern and southern populations had diverged. If climatic conditions were changing once again, then perhaps this would show up in the timing of diapause. The first thing the couple did was go back to look at their old data, to see if it contained any information that they hadn't previously noticed.

"There it was," Holzapfel told me. "Just hitting you right in the eye."

When an animal changes its routine by, say, laying its eggs earlier or going into hibernation later, there are a number of possible explanations. One is that the change reflects an innate flexibility; as conditions vary, the animal is able to adjust its behavior in response. Biologists call such flexibility "phenotypic plasticity," and it is key to the survival of most species. Another possibility is that the shift represents something deeper and more permanent—an actual rearrangement of the organism's genetic code.

In the years since they established their lab, Bradshaw and Holzapfel have collected mosquito larvae from all over the eastern United States and much of Canada. The couple used to do the collecting themselves, driving across the country in a van equipped with a makeshift bed for their daughter and a miniature lab for sorting, labeling, and storing the thousands of specimens they would gather. Nowadays, they more often send out their graduate students, who, instead of driving, are likely to fly. (Getting through airport security with a backpack full of mosquito larvae is a process that, the students have learned, can take half a day.)

Every subpopulation exhibits a range of light responses; Bradshaw and Holzapfel define critical photoperiod as the point at which 50 percent of the mosquitoes in a sample have switched from active development to diapause. Each time they collect a new batch of insects, they put the larvae in petri dishes and place the dishes in the controlled-environment light boxes, which they call Mosquito Hiltons. Then they test the larvae for their critical photoperiod, and record the results.

When Bradshaw and Holzapfel went back to their files, they looked for populations that they had tested at least twice. One of these was from a wetland called Horse Cove, in Macon County, North Carolina. In 1972, when the couple had collected mosquitoes for the first time from Horse Cove, their files showed, the larvae's critical photoperiod was fourteen hours and twenty-one minutes. They collected a second batch of mosquitoes from the same spot in 1996. By that point, the insects' critical photoperiod had dropped to thirteen hours and fifty-three minutes. All told, Bradshaw and Holzapfel found that in their files they had comparative data on ten different subpopulations—two in Florida, three in North Carolina, two in New Jersey, and one each in Alabama, Maine, and Ontario. In every single case, the critical photoperiod had declined over time. Also, their data showed that the farther north you went, the stronger the

effect; a regression analysis revealed that the critical photoperiod of mosquitoes living at fifty degrees north latitude had
declined by more than thirty-five minutes, corresponding to
a delay in diapause of nearly nine days.

In a different mosquito, this shift could be an instance of
the kind of plasticity that allows organisms to cope with
varying conditions. But in *Wyeomyia smithii*, there is no
flexibility when it comes to timing the onset of diapause.
Warm or cold, all the insect can do is read light. Bradshaw
and Holzapfel knew therefore that the change they were
seeing must be genetic. As the climate had warmed, those
mosquitoes that had remained active until later in the fall
had enjoyed a selective advantage, presumably because
they had been able to store a few more days' worth of
resources for the winter, and they had passed this advantage
on to their offspring, and so on. In December 2001,
Bradshaw and Holzapfel published their findings in the

The critical photoperiod for Wyeomyia smithii *has declined markedly over
time. Changes are most dramatic at higher latitudes. Credit: After W.
Bradshaw and C. Holzapfel, PNAS, vol. 98 (2001).*

Proceedings of the National Academy of Sciences. By doing so, they became the first researchers to demonstrate that global warming had begun to drive evolution.

The Monteverde Cloud Forest sits astride the Cordillera de Tilarán, or Tilarán Mountains, in north-central Costa Rica. The rugged terrain in combination with the trade winds that blow off the Caribbean Sea make the region unusually diverse; in an area of less than two hundred and fifty square miles, there are seven "life zones," each with its own distinctive type of vegetation. The cloud forest is surrounded on all sides by land, yet, ecologically speaking, it is an island and, as is often the case with islands, it displays a high degree of endemism, or biological specificity. Fully 10 percent of Monteverdean flora, for example, are believed to be unique to the area.

The most famous of Monteverde's endemic species is—or at least was—a small toad. Known colloquially as the golden toad, it was officially discovered by a biologist from the University of Southern California named Jay Savage. Savage had heard tell of the toad from a group of Quakers who had settled at the edge of the forest; still, when he came across it for the first time, on May 14, 1964, at the top of a high mountain ridge, his reaction, he would later recall, was "one of disbelief." Most toads are dull brown, grayish green, or olive; this one was a flaming shade of tangerine. Savage named the new species *Bufo periglenes*, from the Greek word meaning bright, and titled his paper on the discovery "An Extra-ordinary New Toad (Bufo) from Costa Rica."

Since the golden toad spent its life underground, emer-ging only in order to reproduce, most of what was subsequently learned about it had to do with sex. The toad was, it was determined, an "explosive breeder"; instead of staking out and defending territory, males simply rushed the first available female and fought for the chance to mount her. ("Amplexus" is the term of art for an amphibian embrace.) Males outnumbered females, in some years by as much as ten to one, a situation that often led bachelors to attack amplectant pairs and form what Savage once described as "writhing masses of toad balls." The eggs of the golden toad, black and tan spheres, were deposited in small pools—puddles, really—often no more than one inch deep. Tadpoles emerged in a matter of days, but required another four or five weeks for metamorphosis. During this period, they were highly dependent on the weather; too much rain and they would be washed down the steep hillsides, too little and their puddles would dry up. Golden toads were never found more than a few miles from the site where Savage originally spotted them, always at the top of a mountain ridge, and always at an altitude of between forty-nine hundred and fifty-six hundred feet.

In the spring of 1987, an American biologist who had come to the cloud forest specifically to study the toads counted fifteen hundred of them in temporary breeding pools. That spring was unusually warm and dry, and most of the pools evaporated before the tadpoles in them had

had time to mature. The following year, only one male was seen at what previously had been the major breeding site. Seven males and two females were seen at a second site a few miles away. The year after that, a search of all spots where the toad had earlier been sighted yielded a solitary male. No golden toad has been seen since, and it is widely assumed that after living its colorful, if secretive, existence for hundreds of thousands of years, *Bufo periglenes* is now extinct.

In April 1999, J. Alan Pounds, who heads the Golden Toad Laboratory for Conservation in the Monteverde Preserve, published a paper in *Nature* on the toad's demise. In it, he linked the toad's extinction, as well as the decline of several other amphibian species, to a shift in precipitation patterns in the cloud forest. In recent years, there has been a significant increase in the number of days with no measurable precipitation, a change that, in turn, is consonant with an increase in the elevation of the cloud cover. In a separate article in the same issue of *Nature*, a group of scientists from Stanford University reported on their efforts to model the future of cloud forests. They predicted that as global CO_2 levels continued to rise, the height of the cloud cover in the Monteverde Preserve and other tropical cloud forests would continue to climb. This, they speculated, would force a growing number of high-altitude species "out of existence."

Climate change – even violent climate change – is itself, of course, part of the natural order. For the earth's flora, the last two million years have been particularly turbulent; in addition to the glacial cycles, there have also been dozens of abrupt climate shifts, like the Younger Dryas.

Thompson Webb III is a paleoecologist who teaches at Brown University. He studies pollen grains and fern spores, in an effort to reconstruct the plant life of previous eras. In the mid-seventies, Webb began to assemble a database of pollen records from lakes all across North America. (When a grain of pollen falls on the ground, it usually oxidizes and disappears; if it is blown onto a body of water, however, it can sink to the bottom and be preserved in the sediment for millennia.) The project took nearly twenty years to complete, and, when it was finally done, it showed how, as the climate of the continent had changed, life had rearranged itself.

A few months after I visited Bill Bradshaw and Chris Holzapfel in Eugene, I went to talk to Webb in Providence. He has an office in the university's geochemistry building, and also a lab, where, on this particular day, one of his research assistants was examining charcoal particles from an ancient forest fire. Webb took some slides from a cabinet and slipped one under the lens of a microscope. Most pollen grains are between twenty and seventy microns in diameter; to be identified, they must be magnified four hundred times. Peering through the eyepiece, I saw a tiny sphere, pocked like a golf ball. Webb told me that what I was looking at was a grain of birch pollen. He replaced the slide, and a second tiny golf ball swam into focus. It was beech pollen, Webb explained, and could be distinguished by a set of three minute grooves. "You see, they're really very different," he said of the two grains.

After a while, we went down the hall to Webb's office. On his computer he called up a program named Pollen Viewer 3.2, and a map of North America circa 19000 B.C. appeared on the screen. Around that time, the ice sheets of the last glaciation reached their maximum extent; the map showed the Laurentide ice sheet covering all of Canada as well as most of New England and the upper Midwest. Because so much water was tied up in the ice, sea levels were some three hundred feet lower than they are now. On the map, Florida appeared as a stubby protuberance, nearly twice as wide as it is today. Webb clicked on "Play." Time began to move forward in thousand-year increments. The ice sheet shrank. A huge lake, known as Lake Agassiz, formed in central Canada and, a few thousand years later, drained. The Great Lakes emerged, and then widened. Around eight thousand years ago, open water finally appeared in Hudson Bay. The bay began to contract as the land around it rebounded from the weight of the ice sheet.

Webb clicked on a pull-down menu that listed the Latin names of dozens of trees and shrubs. He chose *Pinus* (pine) and again hit "Play." Dark-green splotches began to move around the continent. Twenty-one thousand years ago, the program showed, pine forests covered the entire Eastern Seaboard south of the ice sheet. Ten thousand years later, pines were concentrated around the Great Lakes, and today pine predominates in the southeastern United States and in western Canada. Webb clicked on *Quercus* (oak), and a similar process began, only *Quercus* moved in a very different pattern from *Pinus*. More clicks for *Fagus* (beech), *Betula* (birch), and *Picea* (spruce). As the earth warmed and the continent emerged from the ice age, each of the tree species migrated, but no two moved in exactly the same way.

"The trick you've got to remember is that climate is multivariate," Webb explained. "The plant species are having to respond both to temperature changes and to moisture changes and to changes in seasonality. It makes a big difference if you have a drier winter versus a drier summer, because some species are more attuned to spring and others to fall. Any current community has a certain mixture. If you start changing the climate, you're changing the temperature, but you're also changing moisture or the timing of the moisture or the amount of snow and, bingo, species are not going to move together. They can't."

Webb pointed out that the warming predicted for the next century is on the same scale as the temperature difference between the last glaciation and today. "You know that's going to give us a very different landscape," he said. I asked what he thought this landscape would look like. He said he didn't know—his central finding, from more than thirty years of research, is that, as the climate changes, species often move in surprising ways. In the short term, which is to say in the remainder of his own life, Webb said that he expected mostly to see disruption.

"We have this strange sense of the evolutionary

hierarchy, that the microorganisms, because they came first, are the most primitive," he told me. "And yet you could argue that this will just give a lot of advantage to the microorganisms of the world, because of their ability to evolve more quickly. To the extent the climate is putting organisms as well as ecosystems under stress, it's opening the opportunities for invasive species on the one hand and disease on the other. I guess I start thinking: Think death."

Any species that is around today, including our own, has already survived catastrophic climate change. The fact that a species has survived such a change, or even many such changes, is no guarantee, however, that it will survive the next one. Consider, for example, the outsized megafauna—seven-hundred-and-fifty-pound saber-toothed cats, elephantine sloths, and fifteen-foot-tall mastodons—that once dominated the North American landscape. These megafauna lived through several glacial cycles, but then something changed, and they nearly all died out at the same time, at the beginning of the Holocene.

Over the past two million years, even as the temperature of the earth has swung wildly, it has always remained within certain limits: The planet has often been colder than today, but rarely warmer, and then only slightly. If the earth continues to warm at the current rate, then by the end of this century temperatures will push beyond the "envelope" of natural climate variability.

Meanwhile, thanks to us, the world today is a very different—and in many ways diminished—place. International trade has introduced exotic pests and competitors; ozone depletion has increased exposure to ultraviolet radiation; and many species have already been very nearly wiped out, or wiped out altogether, by overhunting and overharvesting. Perhaps most significantly, human activity, in the form of farms and cities and subdivisions and mines and logging operations and parking lots, has steadily reduced the amount of available habitat. G. Russell Coope is a visiting professor in the geography department at the University of London and one of the world's leading authorities on ancient beetles. He has shown that, under the pressure of climate change, insects have migrated tremendous distances; for example, *Tachinus caelatus*, a small, dullish-brown beetle common in England during the cold periods of the Pleistocene, today can be found only some five thousand miles away, in the mountains west of Ulan Bator, in Mongolia. But Coope questions whether such long-distance migrations are practical in a fragmented landscape like today's. Many organisms now live in the functional equivalent of "oceanic islands or remote mountain tops," he has written. "Certainly, our knowledge of their past response may be of little value in predicting any future reactions to climate change, since we have imposed totally new restrictions on their mobility; we have inconveniently moved the goal posts and set up a ball game with totally new rules."

A few years ago, nineteen biologists from around the world set out to give, in their words, a "first pass" estimate of the extinction risk posed by global warming. They assembled data on eleven hundred species of plants and animals from sample regions covering roughly a fifth of the earth's surface. Then they established the species' current ranges, based on climate variables such as temperature and rainfall. Finally, they calculated how much of the species' "climate envelope" would be left under different warming scenarios. The results of this effort were published in *Nature* in 2004. Using a midrange projection of temperature rise, the biologists concluded that, if the species in the sample regions could be assumed to be highly mobile, then fully 15 percent of them would be "committed to extinction" by the middle of this century, and, if they proved to be basically stationary, an extraordinary 37 per cent of them would be.

The Mountain Ringlet (*Erebia epiphron*) is a dun-colored butterfly with orange and black spots that curl along the edges of its rounded wings. Mountain Ringlets feed on a coarse, tufted grass known as matgrass, overwinter as larvae, and as adults have an extremely brief lifespan—perhaps as short as one or two days. A montane, or mountain species, it is found only at elevations above a thousand feet in the Scottish Highlands, and farther south, in Britain's Lake District, only above fifteen hundred feet.

Together with a colleague from the University of York, Chris Thomas has for the last few years been monitoring the Mountain Ringlet, along with three other species of butterfly—the Scotch Argus (*Erebia aethiops*), the Large Heath (*Coenonympha tullia*) and the Northern Brown

Argus (*Aricia artaxerxes*)—whose ranges are similarly confined to a few locations in northern England and Scotland.
In the summer of 2004, researchers for the project visited
nearly six hundred sites where these "specialist" species had
been spotted in the past, and the following summer they
repeated the process. Documenting a species' contraction
is more difficult than documenting its expansion—is it
really gone, or did someone just miss it?—but preliminary
evidence suggests that the butterflies are already disappearing from lower elevation, and therefore warmer, sites.
When I went to visit Thomas, he was getting ready to take
his family to Scotland on vacation, and was planning to
recheck some of the sites. "It's a bit of a busman's holiday,"
he confessed.

As we were wandering around his yard in search of
Commas, I asked Thomas, who was the lead author of the
extinction study, how he felt about the changes he was
seeing. He told me that he found the opportunities for
study presented by climate change to be exciting.

"Ecology for a very long time has been trying to explain
why species have the distribution that they do, why a
species can survive here and not over there, why some
species have small distributions and others have broad
ones," he said. "And the problem that we have always
had is that distributions have been rather static. We
couldn't actually see the process of range boundaries
changing taking place, or see what was driving those
changes. Once everything starts moving, we can begin
to understand: is it a climatic determinant, or is it mainly

other things, like interactions with other species? And, of course, if you think of the history of the last million years, we now have the opportunity to try and understand how things might have responded in the past. It's extremely interesting, the prospect of everything changing its distribution, and new mixtures of species from around the world starting to form and produce new biological communities—extremely interesting from a purely academic point of view.

"On the other hand, given our conclusions about possible extinctions, it is, to me personally, a serious concern," he went on. "If we are in the situation where a quarter of the terrestrial species might be at risk of extinction from climate change—people often use the phrase 'being like canaries'—if we've changed our biological system to such an extent, then we do have to get worried about whether the services that are provided by natural ecosystems are going to continue. Ultimately, all of the crops we grow are biological species; all the diseases we have are biological species; all the disease vectors are biological species. If there is this overwhelming evidence that species are changing their distributions, we're going to have to expect exactly the same for crops and pests and diseases. Part of it simply is we've got one planet, and we are heading it in a direction that, quite fundamentally, we don't know what the consequences are going to be."

Part II

MAN

Chapter 5

THE CURSE OF AKKAD

THE WORLD'S FIRST empire was established forty-
three hundred years ago, between the Tigris and
Euphrates Rivers. The details of its founding, by Sargon
of Akkad, have come down to us in a form somewhere
between history and myth. Sargon—Sharru-kin, in the
Akkadian language—means "true king"; almost certainly,
though, he was a usurper. As a baby, Sargon was said to
have been discovered, Moses-like, floating in a basket.
Later, he became cupbearer to the ruler of Kish, one of
ancient Babylonia's most powerful cities. Sargon dreamed
that his master, Ur-Zababa, was about to be drowned by
the goddess Inanna in a river of blood. Hearing about the
dream, Ur-Zababa decided to have Sargon eliminated.
How this plan failed is unknown; no text relating the end
of the story has ever been found.

Until Sargon's reign, Babylonian cities like Kish, and
also Ur and Uruk and Umma, functioned as independent
city-states. Sometimes they formed brief alliances—cunei-
form tablets attest to strategic marriages celebrated and
diplomatic gifts exchanged—but mostly they seem to have

been at war with one another. Sargon first subdued Babylonia's fractious cities, then went on to conquer, or at least sack, lands like Elam, in present-day Iran. He presided over his empire from the city of Akkad, the ruins of which are believed to lie south of Baghdad. It was written that "daily five thousand four hundred men ate at his presence," meaning, presumably, that he maintained a huge standing army. Eventually, Akkadian hegemony extended as far as the Khabur plains, in northeastern Syria, an area prized for its grain production. Sargon came to be known as "king of the world"; later, one of his descendants enlarged this title to "king of the four corners of the universe."

Akkadian rule was highly centralized, and in this way anticipated the administrative logic of empires to come. The Akkadians levied taxes, then used the proceeds to support a vast network of local bureaucrats. They introduced standardized weights and measures—the *gur* equalled roughly three hundred liters—and imposed a uniform dating system, under which each year was assigned the name of a major event that had recently occurred: for instance, "the year that Sargon destroyed the city of Mari." Such was the level of systematization that even the shape and the layout of accounting tablets were imperially prescribed. Akkad's wealth was reflected in, among other things, its artwork, the refinement and naturalism of which were unprecedented.

Sargon ruled, supposedly, for fifty-six years. He was succeeded by his two sons, who reigned for a total of twenty-four years, and then by a grandson, Naram-sin, who declared himself a god. Naram-sin was, in turn, succeeded by his son. Then, suddenly, Akkad collapsed. During one three-year period, four men each, briefly, claimed the throne. "Who was king? Who was not king?" the register known as the Sumerian King List asks, in what may be the first recorded instance of political irony.

The lamentation "The Curse of Akkad" was written within a century of the empire's fall. It attributes Akkad's demise to an outrage against the gods. Angered by a pair of inauspicious oracles, Naram-sin plunders the temple of Enlil, the god of wind and storms, who, in retaliation, decides to destroy both him and his people:

> For the first time since cities were built and founded,
> The great agricultural tracts produced no grain,
> The inundated tracts produced no fish,
> The irrigated orchards produced neither syrup nor
>   wine,
> The gathered clouds did not rain, the *masgurum* did not
>   grow.
> At that time, one shekel's worth of oil was only one-half
>   quart,
> One shekel's worth of grain was only one-half quart . . .
> These sold at such prices in the markets of all the cities!
> He who slept on the roof, died on the roof,
> He who slept in the house, had no burial,
> People were flailing at themselves from hunger.

For many years, the events described in "The Curse of Akkad" were thought, like the details of Sargon's birth, to be purely fictional.

In 1978, after scanning a set of maps at Yale's Sterling Memorial Library, a university archaeologist named Harvey Weiss spotted a promising-looking mound at the confluence of two dry riverbeds in the Khabur plains, near the Iraqi border. He approached the Syrian government for permission to excavate the mound, and, somewhat to his surprise, it was almost immediately granted. Soon, he had uncovered a lost city, which in ancient times was known as Shekhna and today is called Tell Leilan.

Over the next ten years, Weiss, working with a team of students and local laborers, proceeded to uncover an acropolis, a crowded residential neighborhood reached by a paved road, and a large block of grain-storage rooms. He found that the residents of Tell Leilan had raised barley and several varieties of wheat, that they had used carts to transport their crops, and that in their writing they had imitated the style of their more sophisticated neighbors to the south. Like most cities in the region at the time, Tell Leilan had a rigidly organized, state-run economy: people received rations—so many liters of barley and so many of oil—based on how old they were and what kind of work they performed. From the time of the Akkadian empire, thousands of similar potsherds were discovered, indicating that residents had received their rations in mass-produced, one-liter vessels. After examining these and other artifacts,

Weiss constructed a time line of the city's history, from its origins as a small farming village (around 5000 B.C.), to its growth into an independent city of some thirty thousand people (2600 B.C.), and on to its reorganization under imperial rule (2300 B.C.).

Wherever Weiss and his team dug, they also encountered a layer of dirt that contained no signs of human habitation. This layer, which was more than three feet deep, corresponded to the years 2200 to 1900 B.C., and it indicated that, around the time of Akkad's fall, Tell Leilan had been completely abandoned. In 1991, Weiss sent soil samples from Tell Leilan to a lab for analysis. The results showed that, around the year 2200 B.C., even the city's earthworms had died out. Eventually, Weiss came to believe that the lifeless soil of Tell Leilan and the end of the Akkadian empire were products of the same phenomenon—a drought so prolonged and so severe that it represented, in his words, an example of "climate change."

Weiss first published his theory, in the journal *Science*, in August 1993. Since then, the list of cultures whose demise has been linked to climate change has continued to grow. They include the Classic Mayan civilization, which collapsed at the height of its development, around A.D. 800; the Tiwanaku civilization, which thrived near Lake Titicaca, in the Andes, for more than a millennium, then disintegrated around A.D. 1100; and the Old Kingdom of Egypt, which collapsed around the same time as the Akkadian empire. (In an account eerily reminiscent of "The Curse of Akkad," the Egyptian sage Ipuwer described the anguish of the period: "Lo, the desert claims the land. Towns are ravaged . . . Food is lacking . . . Ladies suffer like maidservants. Lo, those who were entombed are cast on high grounds.") In each of these cases, what began as a provocative hypothesis has, as new information has emerged, come to seem more and more compelling. For example, the notion that Mayan civilization had been undermined by climate change was first proposed in the late 1980s, at which point there was little climatological evidence to support it. Then, in the mid-1990s, American scientists studying sediment cores from Lake Chichancanab, in north-central Yucatán, reported that precipitation patterns in the region had indeed shifted during the ninth and tenth centuries, and that this shift had led to periods of prolonged drought. More recently, a group of researchers examining ocean-sediment cores collected off the coast of Venezuela produced an even more detailed record of rainfall in the area. They found that the region experienced a series of severe, "multiyear drought events" beginning around A.D. 750. The collapse of the Classic Mayan civilization, which has been described as "a demographic disaster as profound as any other in human history," is thought to have cost several million lives.

The climate shifts that affected past cultures predate industrialization by hundreds—or, in some cases, thousands—of years. They reflect the climate system's innate variability and could not have been foreseen by the societies that experienced them. Caught by surprise, the Akkadians made sense of their suffering as divine retribution. The climate shifts predicted for the coming century, by contrast, are attributable to forces whose causes we know and whose magnitude we will determine.

The Goddard Institute for Space Studies, or GISS, is situated just south of Columbia University's main campus, at the corner of Broadway and West 112th Street. The institute is not well marked, but most New Yorkers would probably recognize the building: its ground floor is home to Tom's Restaurant, the coffee shop made famous by *Seinfeld*.

GISS, an outpost of NASA, started out, forty-five years ago, as a planetary-research center; today, its major function is making climate forecasts. GISS employs about a hundred and fifty people, many of whom spend their days working on calculations that may—or may not—end up being incorporated in the institute's climate model. Some work on algorithms that describe the behavior of the atmosphere, some on the behavior of the oceans, some on vegetation, some on clouds, and some on making sure that all these algorithms, when they are combined, produce results that seem consistent with the real world. (Once, when some refinements were made to the model, rain nearly stopped falling over the world's rainforests.) The latest version of the GISS model, called ModelE, consists of 125,000 lines of computer code.

GISS's director, James Hansen, occupies a spacious, almost comically cluttered office on the institute's seventh floor. (I must have expressed some uneasiness the first time

I visited him, because the following day I received an e-mail assuring me that the office was "a lot better organized than it used to be.") Hansen, who is sixty-three, is a spare man with a lean face and a fringe of brown hair. Although he has probably done as much to publicize the dangers of global warming as any other scientist, in person he is reticent almost to the point of shyness. When I asked him how he had come to play such a prominent role, he just shrugged. "Circumstances," he said.

Hansen first became interested in climate change in the mid-1970s. Under the direction of James Van Allen (for whom the Van Allen radiation belts are named), he had written his doctoral dissertation on the climate of Venus. In it, he had proposed that the planet, which has an average surface temperature of 876 degrees Fahrenheit, was kept warm by a smoggy haze; soon afterward, a space probe showed that Venus was actually insulated by an atmosphere that consists of 96 percent carbon dioxide. When solid data began to show what was happening to greenhouse gas levels on Earth, Hansen became, in his words, "captivated." He decided that a planet whose atmosphere could change in the course of a human lifetime was more interesting than one that was going to continue, for all intents and purposes, to broil away forever. A group of scientists at NASA had put together a computer program to try to improve weather forecasting using satellite data. Hansen and a team of half a dozen other researchers set out to modify it, in order to make longer-range forecasts about what would happen to global temperatures as greenhouse

gases continued to accumulate. The project, which resulted in the first version of the GISS climate model, took
nearly seven years to complete.

At that time, there was little empirical evidence to
support the notion that the earth was warming. Instrumental temperature records go back, in a consistent fash-
ion, only to the mid-nineteenth century. They show that
average global temperatures rose through the first half of
the twentieth century, then dipped in the 1950s and '60s.
Nevertheless, by the early 1980s Hansen had gained enough confidence in his model to begin to make a series of
increasingly audacious predictions. In 1981, he forecast that
"carbon dioxide warming should emerge from the noise of
natural climate variability" around the year 2000. During
the exceptionally hot summer of 1988, he appeared before
a Senate subcommittee and announced that he was "99
percent" sure that "global warming is affecting our planet
now." And in the summer of 1990 he offered to bet a
roomful of fellow scientists a hundred dollars that either
that year or one of the following two years would be the
warmest on record. To qualify, the year would have to set
a record not only for land temperatures but also for sea-
surface temperatures and for temperatures in the lower
atmosphere. Hansen won the bet in six months.

Like all climate models, GISS's divides the world into a
series of boxes. Thirty-three hundred and twelve boxes
cover the earth's surface, and this pattern is repeated
twenty times moving up through the atmosphere, so that

Climate models divide the world into a series of boxes. Credit: Global
Warming: The Complete Briefing, *Cambridge University Press.*

the whole arrangement might be thought of as a set of
enormous checkerboards stacked on top of one another.
Each box represents an area of four degrees latitude by five
degrees longitude. (The height of the box varies depending
on altitude.) In the real world, of course, such a large area
would have an incalculable number of features; in the
world of the model, features such as lakes and forests and,
indeed, whole mountain ranges are reduced to a limited set
of properties, which are then expressed as numerical

approximations. Time in this grid-world moves ahead for the most part in discrete, half-hour intervals, meaning that a new set of calculations is performed for each box for every thirty minutes that is supposed to have elapsed in actuality. Depending on what part of the globe a box represents, these calculations may involve dozens of different algorithms, so that a model run that is supposed to simulate climate conditions over the next hundred years involves more than a quadrillion separate operations. A single run of the GISS model, done on a supercomputer, usually takes about a month.

Very broadly speaking, there are two types of equations that go into a climate model. The first group expresses fundamental physical principles, like the conservation of energy and the law of gravity. The second group describes—the term of art is "parameterize"—patterns and interactions that have been observed in nature but may be only partly understood, or processes that occur on a small scale and have to be averaged out over huge spaces. Here, for example, is a tiny piece of ModelE, written in the computer language FORTRAN, which deals with the formation of clouds:

```
C**** COMPUTE THE AUTOCONVERSION RATE OF CLOUD WATER TO
PRECIPITATION
    RHO=1.E5*PL(L)/(RGAS*TL(L))
    TEM=RHO*WMX(L)/(WCONST*FCLD+1.E-20)
    IF(LHX.EQ.LHS) TEM=RHO*WMX(L)/(WMUI*FCLD+1.E=20)
    TEM=TEM*TEM
    IF(TEM.GT.10.) TEM=10.
    CM1=CM0
    IF(BANDF) CM1=CM0*CBF
    IF(LHX.EQ.LHS) CM1=CM0
```

```
CM=CM1*(1.-1./EXP(TEM*TEM))+1.*100.*(PREBAR(L + 1) +
* PRECNVL(L+1)*BYDTsrc)
IF(CM.GT.BYDTsrc) CM=BYDTsrc
PREP(L)=WMX(L)*CM
END IF
C**** FORM CLOUDS ONLY IF RH GT RH00
219 IF(RH1(L).LT.RH00(L)) GO TO 220
C**** COMPUTE THE CONVERGENCE OF AVAILABLE LATENT HEAT
SQ(L)=LHX*QSATL(L)*DQSATDT(TL(L),LHX)*BYSHA
TEM=-LHX*DPDT(L)/PL(L)
QCONV=LHX*AQ(L)-RH(L)*SQ(L)*SHA*PLK(L)*ATH(L)
* -TEM*QSATL(L)*RH(L)
IF(QCONV.LE.0.0.AND.WMX(L).LE.0) GO TO 220
C**** COMPUTE EVAPORATION OF RAIN WATER, ER
RHN=RHF(L)
IF(RHF(L).GT.RH(L)) RHN=RH(L)
```

All climate models treat the laws of physics in the same way, but, since they parameterize phenomena like cloud formation differently, they come up with different results. Also, because the real-world forces influencing the climate are so numerous, different models tend, like medical students, to specialize. GISS's model specializes in the behavior of the atmosphere; other models in the behavior of the oceans; and still others in the behavior of land surfaces and ice sheets.

One rainy November afternoon, I attended a meeting at GISS that brought together members of the institute's modeling team. When I arrived, about twenty men and five women were sitting in battered chairs in a conference room across from Hansen's office. At that particular moment, the institute was performing a series of runs for the U.N. Intergovernmental Panel on Climate Change. The runs were overdue, and apparently the IPCC was getting impatient. Hansen flashed a series of charts on a screen on the wall summarizing some of the results obtained so far.

The obvious difficulty in verifying any particular climate model or climate-model run is the prospective nature of the results. For this reason, models are often run into the past, to see how well they reproduce trends that have already been observed. Hansen told the group that he was pleased with how ModelE had reproduced the aftermath of the eruption of Mount Pinatubo, in the Philippines, which took place in June 1991. Volcanic eruptions release huge quantities of sulfur dioxide—Pinatubo produced some twenty million tons of the gas—which, once in the stratosphere, condenses into tiny sulfate droplets. These droplets, or aerosols, tend to cool the earth by reflecting sunlight back into space. Man-made aerosols, produced by burning coal, oil, and biomass, also reflect sunlight and are a countervailing force to greenhouse warming, albeit one with serious health consequences of its own. (The impact of man-made aerosols is difficult to quantify; without it, however, the earth almost certainly would have warmed even faster than it has.) The cooling effect of aerosols lasts only as long as the droplets remain suspended in the atmosphere. In 1992, following the Pinatubo eruption, global temperatures, which had been rising sharply, fell by half a degree. Then they began to climb again. ModelE had succeeded in simulating this effect to within nine hundredths of a degree. "That's a pretty nice test," Hansen observed laconically.

★　　★　　★

One day, when I was talking to Hansen in his cluttered office, he pulled a pair of photographs out of his briefcase. The first showed a chubby-faced five-year-old girl holding some miniature Christmas-tree lights in front of an even chubbier-faced five-month-old baby. The girl, Hansen told me, was his granddaughter Sophie and the boy was his new grandson, Connor. The caption on the first picture read, "Sophie explains greenhouse warming." The caption on the second photograph, which showed the baby smiling gleefully, read, "Connor gets it."

When modelers talk about what drives the climate, they focus on what they call "forcings." A forcing is any ongoing process or discrete event that alters the energy of the system. Examples of natural forcings include, in addition to volcanic eruptions, periodic shifts in the earth's orbit and changes in the sun's output, like those linked to sunspots. Many climate shifts of the past have no known forcing associated with them; for instance, no one is certain what brought about the so-called Little Ice Age, the cool period that lasted in Europe from around 1500 to 1850. A very large forcing, meanwhile, should produce a com-mensurately large—and obvious—effect. One GISS scientist put it to me this way: "If the sun went supernova, there's no question that we could model what would happen."

Adding carbon dioxide, or any other greenhouse gas, to the atmosphere by, say, burning fossil fuels or leveling forests is, in the language of climate science, an anthropogenic forcing. Since preindustrial times, the concentration of CO_2 in the atmosphere has risen by roughly a third, from 280 to 378 parts per million. During the same period, the concentration of methane has more than doubled, from .78 to 1.76 parts per million. Scientists measure forcings in terms of watts per square meter, or w/m^2, by which they mean that a certain number of watts have been added (or, in the case of a negative forcing, like aerosols, subtracted) for every single square meter of the earth's surface. The size of the greenhouse forcing is estimated, at this point, to be 2.5 w/m^2. A miniature Christmas light gives off about four tenths of a watt of energy, mostly in the form of heat, so that, in effect (as Sophie supposedly explained to Connor), we have covered the earth with tiny bulbs, six for every square meter. These bulbs are burning twenty-four hours a day, seven days a week, year in and year out.

If greenhouse gases were held constant at today's levels, it is estimated that it would take several decades for the full impact of the forcing that is already in place to be felt. This is because raising the earth's temperature involves not only warming the air and the surface of the land but also melting sea ice, liquefying glaciers, and, most significant, heating the oceans, all processes that require tremendous amounts of energy. (Imagine trying to thaw a gallon of ice cream or warm a pot of water using an Easy-Bake oven.) The delay that is built into the system is, in a certain sense, fortunate. It enables us, with the help of climate models, to foresee what is coming and therefore to prepare for it. But in another sense it is clearly disastrous, because it allows us to keep adding CO_2 to the atmosphere while fobbing the impacts off on our children and grandchildren.

There are two ways to operate a climate model. In the first, which is known as a transient run, greenhouse gases are slowly added to the simulated atmosphere—just as they would be to the real atmosphere—and the model forecasts what the effect of these additions will be at any given moment. In the second, greenhouse gases are added to the atmosphere all at once, and the model is run at these new levels until the climate has fully adjusted to the forcing by reaching a new equilibrium. (Not surprisingly, this is known as an equilibrium run.)

For doubled CO_2, equilibrium runs of the GISS model predict that average global temperatures will rise by 4.9 degrees Fahrenheit. Only about a third of this increase is directly attributable to higher greenhouse gas levels. The rest is a result of indirect effects, like the melting of sea ice, which allows the earth to absorb more heat. The most significant indirect effect is known as the "water-vapor feedback." Since warmer air holds more moisture, higher temperatures are expected to produce an atmosphere containing more water vapor, which is itself a greenhouse gas. GISS's forecast is on the low end of the most recent projections for doubled CO_2; the Hadley Centre model, which is run by the British Met Office, predicts that under these conditions, the eventual temperature rise will be 6.3 degrees Fahrenheit, while Japan's National Institute for Environmental Studies predicts that it will be 7.7 degrees.

In the context of ordinary life, a warming of 4.9, or even of 7.7, degrees may not seem like much to worry about. On the dreary November day I attended the GISS modeling meeting, the temperature in Central Park was fifty-two degrees at seven A.M., and by two P.M. had reached sixty degrees. In the course of a normal summer's day, air temperatures routinely rise by fifteen degrees or more. Average global temperatures, however, have practically nothing to do with ordinary life. This is perhaps best illustrated by the ups and downs of climate history. The so-called Last Glacial Maximum—the point during the most recent glaciation when the ice sheets reached their maximum extent —occurred about twenty thousand years ago. At that time, the Laurentide ice sheet reached deep into what is now the northeastern and midwestern United States, and sea levels were so low that Siberia and Alaska were connected by a land bridge nearly a thousand miles wide. During the Last Glacial Maximum, average global temperatures were only about ten degrees colder than they are today. It is worth noting that the total forcing that ended that ice age is estimated to have been just six and a half watts per square meter.

David Rind is a climate scientist who has worked at GISS since 1978. Rind acts as a troubleshooter for the institute's model, scanning reams of numbers known as diagnostics, trying to catch problems, and he also works with what is known as the GISS Climate Impacts Group. (His office, like Hansen's, is filled with dusty piles of computer printouts.) Although higher temperatures are the most predictable result of increased CO_2, other, second-order consequences—rising sea levels, changes in vegetation, loss of snow cover—are likely to be just as significant. Rind's particular interest is how CO_2 levels will affect water supplies, because, as he put it to me, "you can't have a plastic version of water."

One afternoon, when I was talking to Rind in his office, he mentioned a visit that President George W. Bush's science adviser, John Marburger III, had paid to GISS a few years earlier. "He said, 'We're really interested in adaptation to climate change,'" Rind recalled. "Well, what does 'adaptation' mean?" He rummaged through one of his many file cabinets and finally pulled out a paper that he had published in the *Journal of Geophysical Research* titled "Potential Evapotranspiration and the Likelihood of Future Drought." In much the same way that wind velocity is measured using the Beaufort scale, water availability is measured using the Palmer Drought Severity Index. Different climate models offer very different predictions about future water availability; in the paper, Rind applied the criteria used in the Palmer index to GISS's model and also to a model operated by the National Oceanic and Atmospheric Administration's Geophysical Fluid Dynamics Laboratory. He found that as carbon dioxide levels rose, the world would begin to experience more and more serious water shortages, starting near the equator and then spreading toward the poles. When he applied the index to the GISS model for doubled CO_2, it showed most of the continental United States would be suffering under severe drought conditions. When he applied the index to the GFDL model, the results were even more dire. Rind created two maps to illustrate these findings. Yellow represented a 40 to 60 percent chance of summertime drought, ochre a 60 to 80 percent chance, and brown an 80 to 100 percent chance. In the first map, showing the GISS results, the Northeast was yellow, the Midwest was ochre, and the Rocky Mountain states and California were brown. In the second, showing the GFDL results, brown covered practically the entire country.

"I gave a talk based on these drought indices out in California to water-resource managers," Rind told me. "And they said, 'Well, if that happens, forget it.' There's just no way they could deal with that."

He went on, "Obviously, if you get drought indices like these, there's no adaptation that's possible. But let's say it's not that severe. What adaptation are we talking about? Adaptation in 2020? Adaptation in 2040? Adaptation in 2060? Because the way the models project this, as global warming gets going, once you've adapted to one decade, you're going to have to change everything the next decade.

"We may say that we're more technologically able than earlier societies. But one thing about climate change is it's potentially geopolitically destabilizing. And we're not only more technologically able; we're more technologically able destructively as well. I think it's impossible to predict what will happen. I guess—though I won't be around to see it—I wouldn't be shocked to find out that by 2100 most things were destroyed." He paused. "That's sort of an extreme view."

On the other side of the Hudson River and slightly to the north of GISS, the Lamont-Doherty Earth Observatory occupies what was once a weekend estate in the town of Palisades, New York. The observatory is an outpost of Columbia University, and it houses, among its collections of natural artifacts, the world's largest assembly of ocean-sediment cores—more than thirteen thousand in all. The cores are kept in steel compartments that look like drawers from a filing cabinet, only longer and much skinnier. Some of the cores are chalky, some are clayey, and some are made up almost entirely of gravel. All can be coaxed to yield up—in one way or another—information about past climates.

Peter deMenocal is a paleoclimatologist who has worked at Lamont-Doherty for fifteen years. DeMenocal is an expert on ocean cores, and also on the climate of the Pliocene, which lasted from roughly five million to two million years ago. Around two and a half million years ago, the earth, which had been warm and relatively ice-free, started to cool down until it entered an era—the Pleistocene—of recurring glaciations. DeMenocal has argued that this transition was a key event in human evolution: right around the time that it occurred, at least two types of hominids—one of which would eventually give rise to modern man—branched off from a single ancestral line. Until quite recently, paleoclimatologists like deMenocal rarely bothered with anything much closer to the present day; the current interglacial—the Holocene—was believed to be too stable to warrant much study. In the mid-nineties, though, deMenocal, motivated by a growing concern over global warming— and a concomitant shift in government research funds— decided to look in detail at some Holocene cores. What he learned about the period, as he put it to me when I visited him at Lamont-Doherty, was that it was "less boring than we had thought."

One way to extract climate data from ocean sediments is to examine the remains of what lived or, perhaps more pertinently, what died and was buried there. The oceans are rich with microscopic creatures known as foraminifera—forams, for short. Forams are tiny, single-celled organisms that construct shells out of calcite. These shells come in a wide range of shapes; viewed under a microscope, some look like tiny sand dollars, others like conch shells, and still others like lumps of dough. There are about thirty planktonic species of foraminifera—which is to say, species that live near the top of the sea—and each thrives at a different water temperature, so that by counting a species' prevalence in a given sample it is possible to estimate how warm (or cold) the ocean was at the time the sediment was formed. When deMenocal used this technique to analyze cores that had been collected off the coast of Mauritania, he found that they contained evidence of recurring cool periods; every fifteen hundred years or so, water temperatures would drop for a few centuries before climbing back up again. (The most recent cool period corresponds to the Little Ice Age, which ended about a century and a half ago.) The cores also showed dramatic changes in precipitation. Until about six thousand years ago, northern Africa was relatively wet—dotted with small lakes. Then it became dry, as it is today. DeMenocal traced the shift to periodic variations in the earth's orbit, which, in a generic sense, are the same forces that trigger ice ages. But orbital changes occur gradually, over thousands of years, and northern Africa appears to have switched from wet to dry all of a sudden. Although no one knows exactly how this happened, it seems, like so many climate events, to have been a function of feedbacks—the less rain the continent got, the less vegetation there was to retain water, and so on until, finally, the system just flipped. The process provides yet more evidence of how a very small forcing sustained over time can produce dramatic results.

"We were kind of surprised by what we found," deMenocal told me about his work on the supposedly stable Holocene. "Actually, more than surprised. It was one of these things where, you know, in life you take certain things for granted, like your neighbor's not going to be an ax murderer. And then you discover your neighbor *is* an ax murderer."

Not long after deMenocal began to think about the Holocene, a brief mention of his work on the climate of Africa appeared in a book produced by *National*

Geographic. On the facing page, there was a piece on Harvey Weiss and his work at Tell Leilan. DeMenocal vividly remembers his reaction. "I thought, Holy cow, that's just amazing!" he told me. "It was one of these cases where I lost sleep that night, I just thought it was such a cool idea."

DeMenocal also recalls his subsequent dismay when he went to learn more. "It struck me that they were calling on this climate-change argument, and I wondered how come I didn't know about it," he said. He looked at the *Science* paper in which Weiss had originally laid out his theory. "First of all, I scanned the list of authors and there was no paleoclimatologist on there," deMenocal said. "So then I started reading through the paper and there basically was no paleoclimatology in it." (The main piece of evidence Weiss adduced for a drought was that Tell Leilan had filled with dust.) The more deMenocal thought about it, the more unconvincing he found the data and the more compelling he found the underlying idea. "I just couldn't leave it alone," he told me. In the summer of 1995, he went with Weiss to Syria to visit Tell Leilan. Subsequently, he decided to do his own study to prove—or disprove—Weiss's theory.

Instead of looking in, or near, the ruined city, de-Menocal focused on the Gulf of Oman, a thousand miles downwind. Dust from the Mesopotamian floodplains, just north of Tell Leilan, contains heavy concentrations of the mineral dolomite, and since arid soil produces more wind-borne dust, deMenocal figured that if there had been a drought of any magnitude it would show up in gulf sediments. "In a wet period, you'd be getting none or very, very low amounts of dolomite, and during a dry period you'd be getting a lot," he explained. He and a graduate student named Heidi Cullen developed a highly sensitive test to detect dolomite, and then Cullen assayed, centimeter by centimeter, a sediment core that had been extracted near where the Gulf of Oman meets the Arabian Sea.

"She started going up through the core," deMenocal told me. "It was like nothing, nothing, nothing, nothing, nothing. Then one day, I think it was a Friday afternoon, she goes, 'Oh, my God.' It was really classic." DeMenocal had thought that the dolomite level, if it were elevated at all, would be modestly higher; instead, it went up by 400 percent. Still, he wasn't satisfied. He decided to have the core reanalyzed using a different marker: the ratio of strontium 86 and strontium 87 isotopes. The same spike showed up. When deMenocal had the core carbon-dated, it turned out that the spike lined up exactly with the period of Tell Leilan's abandonment.

Tell Leilan was never an easy place to live. Much like, say, western Kansas today, the Khabur plains received enough annual rainfall—about seventeen inches—to support cereal crops, but not enough to grow much else. "Year-to-year variations were a real threat, and so they obviously needed to have grain storage and to have ways to buffer themselves," deMenocal observed. "One generation would tell the next, 'Look, there are these things that happen that you've got to be prepared for.' And they were good at that. They could manage that. They were there for hundreds of years."

He went on, "The thing they couldn't prepare for was the same thing that we won't prepare for, because in their case they didn't know about it and because in our case the political system can't listen to it. And that is that the climate system has much greater things in store for us than we think."

Shortly before Christmas 2004, Harvey Weiss gave a lunchtime lecture at Yale's Institute for Biospheric Studies. The title was "What Happened in the Holocene," which, as Weiss explained, was an allusion to a famous archaeology text by V. Gordon Childe, titled *What Happened in History*. The talk brought together archaeological and paleoclimatic records from the Near East over the last ten thousand years.

Weiss, who is sixty years old, has thinning gray hair, wire-rimmed glasses, and an excitable manner. He had prepared for his audience—mostly Yale professors and graduate students—a handout with a time line of Mesopotamian history. Key cultural events appeared in black ink, key climatological ones in red. The two alternated in a rhythmic cycle of disaster and innovation. Around 6200 B.C., a severe global cold snap—red ink—produced aridity in the Near East. (The cause of the cold snap is believed to have been a catastrophic flood that emptied an enormous glacial lake—Lake Agassiz—into the North

Atlantic.) Right around the same time—black ink—farming villages in northern Mesopotamia were abandoned, while in central and southern Mesopotamia the art of irrigation was invented. Three thousand years later, there was another cold snap, after which settlements in northern Mesopotamia once again were deserted. The most recent red event, in 2200 B.C., was followed by the dissolution of the Old Kingdom in Egypt, the abandonment of villages in ancient Palestine, and the fall of Akkad. Toward the end of his talk, Weiss, using PowerPoint, displayed some photographs from the excavation at Tell Leilan. One showed the wall of a building—probably intended for administrative offices—that had been under construction when the rain stopped. The wall was made from blocks of basalt topped by rows of mud bricks. The bricks gave out abruptly, as if construction had ceased from one day to the next.

The monochromatic sort of history that most of us grew up with did not allow for events like the drought that destroyed Tell Leilan. Civilizations fell, we were taught, because of wars or barbarian invasions or political unrest. (Another famous text by Childe bears the exemplary title *Man Makes Himself*.) Adding red to the time line points up the deep contingency of the whole enterprise. Civilization goes back, at the most, ten thousand years, even though, evolutionarily speaking, modern man has been around for at least ten times that long. The climate of the Holocene was not boring, but it was dull enough to allow people to sit still. It was only after the immense climatic shifts of the glacial epoch had run their course that agriculture and writing finally emerged.

Nowhere else does the archaeological record go back so far or in such detail as in the Near East. But similar red-and-black chronologies can now be drawn up for many other parts of the world: the Indus Valley, where, some four thousand years ago, the Harappan civilization suffered a decline after a change in monsoon patterns; the Andes, where, fourteen hundred years ago, the Moche abandoned their cities in a period of diminished rainfall; and even the United States, where the arrival of the English colonists on Roanoke Island, in 1587, coincided with a severe regional drought. (By the time English ships returned to resupply the colonists, three years later, not a single one was left.) At the height of the Mayan civilization, population density was five hundred per square mile, higher than it is in most parts of the United States today. Two hundred years later, most Mayan territory had been completely depopulated. You can argue that man through culture creates stability, or you can argue, just as plausibly, that stability is for culture an essential precondition.

After the lecture, I walked with Weiss back to his office, which is near the center of the Yale campus, in the Hall of Graduate Studies. In 2004, Weiss decided to suspend excavation at Tell Leilan. The site lies only fifty miles from the Iraqi border, and, owing to the uncertainties of the war, it seemed like the wrong sort of place to bring graduate students. When I visited, Weiss had just returned from a trip to Damascus, where he had gone to pay the guards who watch over the site when he isn't there. While he was away from his office, its contents had been piled up in a corner by repairmen who had come to fix some pipes. Weiss considered the piles disconsolately, then unlocked a door at the back of the room.

The door led to a second room, much larger than the first. It was set up like a library, except that instead of books the shelves were stacked with hundreds of cardboard boxes. Each box contained fragments of broken pottery from Tell Leilan. Some were painted, others were incised with intricate designs, and still others were barely distinguishable from pebbles. Every fragment had been inscribed with a number, indicating its provenance.

I asked what he thought life in Tell Leilan had been like. Weiss told me that that was a "corny question," so I asked him about the city's abandonment. "Nothing allows you to go beyond the third or fourth year of a drought, and by the fifth or sixth year you're probably gone," he observed. "You've given up hope for the rain, which is exactly what they wrote in 'The Curse of Akkad.'" I said I would like to see something that might have been used in Tell Leilan's last days. Swearing softly, Weiss searched through the rows until he finally found one particular box. It held several potsherds that appeared to have come from identical bowls. They were made from a greenish-colored clay, had been thrown on a wheel, and had no decoration. Intact, the bowls had held about a liter, and Weiss explained that they had been used to mete out rations—probably wheat or barley—to the workers of Tell Leilan. He passed me one of the fragments. I held it in my hand for a moment and tried to imagine the last Akkadian who had touched it. Then I passed it back.

Chapter 6

FLOATING HOUSES

IN FEBRUARY 2003, a series of ads on the theme of inundation began appearing on Dutch TV. The ads were sponsored by the Netherlands' Ministry of Transport, Public Works, and Water Management, and they featured a celebrity weatherman named Peter Timofeeff. In one commercial, Timofeeff, who looks a bit like Albert Brooks and a bit like Gene Shalit, sat relaxing on the shore in a folding chair. "Sea level is rising," he announced, as waves started creeping up the beach. He continued to sit and talk even as a boy who had been building a sand castle abandoned it in panic. At the end of the ad, Timofeeff, still seated, was immersed in water up to his waist. In another commercial, Timofeeff was shown wearing a business suit and standing by a bathtub. "These are our rivers," he explained, climbing into the tub and turning on the shower full blast. "The climate is changing. It will rain more often, and more heavily." Water filled the tub and spilled over the sides. It dripped through the floorboards, onto the head of his screeching wife below. "We should give the water more space and widen the rivers," he advised, calmly reaching for a towel.

Both the beach chair and the shower ads were part of a public-service campaign titled, somewhat ambiguously, *"Nederland Leeft Met Water"* ("The Netherlands Lives with Water"), which also included radio spots, free tote bags, and newspaper announcements drawn in the form of cartoons. Its tone was consistently lighthearted—other commercials showed Timofeeff trying to start a motorboat in a cow pasture and digging a duck pond in his back-yard—either in spite of the fact that, or maybe precisely because, its message for the Dutch was so devastating.

Fully a quarter of the Netherlands lies below sea level, on land wrested from either the North Sea or the Rhine or the Meuse Rivers, or one of the hundreds of natural lakes that once dotted the countryside. Another quarter, while slightly higher, is still low enough that, in the natural course of events, it would regularly be flooded. What has made this arrangement possible is the world's most so-phisticated water management system, which, according to government figures, comprises 150 miles of dunes, 260 miles of sea dikes, 850 miles of river dikes, 610 miles of lake dikes, and 8,000 miles of canal dikes, not to mention countless pumps, holding ponds, and windmills.

Historically, whenever flooding has occurred, the Dutch response has been either to reinforce the dikes or to add new ones. In 1916, for example, after the defenses gave out along an inlet of the North Sea known as the Zuiderzee, the Dutch dammed up the Zuiderzee, creating an artificial lake as large as Los Angeles. In 1953, storms overwhelmed the dikes in the province of Zeeland, killing 1,835 people. Immediately afterward, the government embarked on a massive, five-and-a-half-billion-dollar construction project known as the Delta Works. (The last phase of the project, the Maeslant barrier, which was finally completed in 1997, is supposed to protect Rotterdam from storm surges with the aid of two moveable arms, each the size of a skyscraper.) People in Holland like to joke, although they are not really joking, "God made the world, but the Dutch made the Netherlands."

"The Netherlands Lives with Water" signals the end of this five-hundred-year project. Looking ahead, the same engineers who built the Maeslant barrier have determined that even such monumental public works projects are no longer adequate. From now on, instead of reclaiming land from the water, the Dutch, they have decided, are going to have to start giving back.

The Nieuwe Merwede looks like a river, but is actually a canal, dug in the 1870s in the delta of the Rhine and Meuse Rivers. It runs on a winding course west from the city of Werkendam until it meets up with another man-made river to form what is known as the Hollandsch Diep, which, in the confusion of the delta, splits again, and eventually empties into the North Sea.

On the north side of the canal, in a pocket-sized national park called Biesbosch, is a nature center, which, at the time that I visited, was running an exhibit on climate change. By way of decoration, large black umbrellas had been hung from the ceiling, and in the background, the soundtrack of a church bell—the traditional Dutch flood warning—tolled periodically. One kid-friendly display allowed visitors to turn a crank and, in effect, drown the countryside. By 2100, the display showed, the Nieuwe Merwede will be running, at peak flows, several feet above the top of the local dikes.

There are several reasons why global warming produces flooding. The first has to do simply with the physics of liquids. As water warms, it expands. In a small body of water, the effect is small; in a big body, it's commensurately larger. Most of the sea level rise predicted for the next hundred years—a total of up to three feet—is purely a function of thermal expansion. (Even if greenhouse gas levels are eventually stabilized, sea levels will continue to rise for several centuries, owing to the oceans' thermal inertia.)

Meanwhile, a warming Earth means changing precipitation patterns. Just as some regions, like the American Midwest, are predicted to suffer from drought, others will experience more—or at least more intense—rainfall. The effect is likely to be particularly punishing in some of the most densely populated regions on Earth, including the Mississippi Delta, the Ganges Delta, and the Thames basin. A study commissioned a few years ago by the British government concluded that under certain conditions, floods of a magnitude now expected no more often than once a century could, by 2080, be occurring in England once every three years. (As it happened, the very week I was in the Netherlands, thirteen people were killed by exceptionally heavy winter storms in Britain and Scandinavia.)

At the Biesbosch nature center, I met up with a water-ministry official named Eelke Turkstra. Turkstra runs a program called *Ruimte voor de Rivier* (Room for the River), and these days his job consists not in building dikes, but in dismantling them. He explained to me that the Dutch were already seeing more rainfall than they used to. Where once the water ministry had planned on peak flows in the Rhine of no more than fifteen thousand cubic meters per second, recently it had been forced to raise that to sixteen thousand cubic meters per second and was already anticipating having to deal with eighteen thousand cubic meters per second. Rising sea levels, meanwhile, were likely to further compound the problem by impeding the flow of the river to the ocean.

"We think in the run of this century, sea levels can rise by sixty centimeters," or just under two feet, Turkstra told me. "When that happens—we're sure that it *will* happen—that makes things very complicated."

From the nature center, we took a car ferry across the Nieuwe Merwede. The area we were driving through was made up entirely of "polders"—land that has been laboriously reclaimed from the water. The polders were shaped like ice trays, with sloping sides and perfectly flat fields along the bottom. Every once in a while, there was a sturdy-looking farmhouse. The whole scene—the level fields, the thatched barns, even the gray clouds sitting on the horizon—could have been borrowed from a painting by Hobbema. All this land, Turkstra said, was destined for inundation. The plan of Room for the River was to buy out the farmers who were living in the polders, and then lower the surrounding dikes. By selectively abandoning rural areas like this one, the water ministry hoped to be able to protect population centers like the nearby city of Gorinchem. The price tag for the project we were looking at had been set at $390 million. Similar projects were under way in other parts of the Netherlands, and still others were in the design phase. Some of the designs had provoked angry, ongoing protests. Surrendering land that people have been living on for decades, in some cases centuries, was, Turkstra acknowledged, bound to cause political problems, but that was precisely the reason that it was important to get started immediately.

"Some people don't get it," he told me as we zipped along. "They think this project is stupid. But I think it's stupid to continue the old way."

When climatologists discuss the hazards of rising greenhouse gas levels, they use the phrase "dangerous anthropogenic interference" or, for short, DAI. The term does not refer to any disaster in particular, although there are, it is generally agreed, a number of scenarios that would fit the bill—climate change dramatic enough to destroy entire ecosystems, for instance, or cause mass extinction or disrupt the world's food supply. The disintegration of one of the planet's remaining ice sheets is often held up as the exemplary catastrophe. The West Antarctic ice sheet is, at this point, the world's only marine ice sheet, meaning that it rests on land that is below sea level. For this reason it is considered particularly vulnerable to collapse. Were the West Antarctic or the Greenland ice sheet to be destroyed, sea levels around the world would rise by at least fifteen feet. Were both ice sheets to disintegrate, global sea levels would rise by thirty-five feet. It could take centuries for either of the ice sheets to disappear entirely, but once disintegration got under way it would start to feed on itself, most likely becoming irreversible. Other catastrophes have similar built-in delays, which follow from the tremendous inertia of the climate system. DAI is therefore understood to refer not to the end of the process—the moment when disaster actually arrives—but to the beginning of it: the point at which its arrival becomes unavoidable.

Exactly what forcing or temperature or level of CO_2 represents DAI is a question of the utmost significance and one that cannot at this point be answered. Policy studies often take 500 parts per million of CO_2—roughly double preindustrial levels—as the threshold. But this figure has at least as much to do with what appears to be a socially feasible goal as with what has been scientifically demonstrated.

In the last decade, a great deal has been discovered about how the climate functions, both through measurements made in real time and through reconstructions of the paleoclimatic record. Just about everything that has been learned—from the observed acceleration of the ice sheets to the inferred history of the thermohaline circulation—has tended to push the level of DAI downward. Many climate scientists now believe that 450 parts per million of CO_2 represents a more objective estimate of danger, while others argue that the threshold is 400 parts per million or even lower.

Probably the most significant of the recent discoveries was made in Antarctica, at a research base known as the Vostok station. Between 1990 and 1998, an 11,775-foot-long ice core was drilled there. Since less snow falls in Antarctica than in Greenland, the layers in an Antarctic core are thinner and the climate information contained in them is less detailed. However, they go back much farther. The Vostok core, which is now stored in pieces in Denver, Grenoble, and on Antarctica, contains a continuous climate record stretching back four full glacial cycles. (As is the case with Greenland cores, temperatures can be ascertained by measuring the isotopic composition of the ice, and the makeup of the atmosphere determined by analyzing tiny bubbles of trapped air.)

What the Vostok record shows is that the planet is already nearly as warm as it has been at any point in the last 420,000 years. A possible consequence of even a four- or five-degree temperature rise—on the low end of projections for the end of this century—is that the world will enter a completely new climate regime, one with which modern humans have no prior experience. When it comes to carbon dioxide, meanwhile, the evidence is even more striking. The Vostok record demonstrates that, at 378 parts

The record from the Vostok core shows that CO_2 levels and temperatures have varied in tandem. Current CO_2 levels are unprecedented in the last 420,000 years. Credit: J.R. Petit et al, Nature, vol. 399 (1999).

per million, current CO_2 levels are unprecedented in recent geological history. (The previous high, of 299 parts per million, was reached around 325,000 years ago). It is believed that the last time carbon dioxide levels were comparable to today's was three and a half million years ago, during what is known as the mid–Pliocene warm period, and it is likely that they have not been much higher since the Eocene, some fifty million years ago. In the Eocene, crocodiles roamed Colorado and sea levels were nearly three hundred feet higher than they are today. A scientist with the National Oceanic and Atmospheric Administration (NOAA) put it to me—only half-jokingly—this way: "It's true that we've had higher CO_2 levels before. But, then, of course, we also had dinosaurs."

The town of Maasbommel is situated about fifty miles east of Biesbosch. It lies on the banks of the River Meuse and is a popular holiday destination; every summer it fills with tourists who have come to go boating or to camp out. Thanks to the risk of flooding, building is restricted along the river, but a few years ago one of the Netherlands' largest construction firms, Dura Vermeer, received permission to turn a former RV park on the banks of the Meuse into a development of "amphibious homes."

The first of the amphibious homes were completed in the fall of 2004, and on a dull winter's day a few months afterward, I went to take a look at them. On my way, I stopped off at Dura Vermeer's headquarters to meet with the company's environmental director, Chris Zevenbergen. In his office, Zevenbergen played for me an animated video on the future of the Netherlands; it showed large chunks of the country gradually being swallowed up by water. It was lunchtime, and after a while his secretary came around carrying a tray of sandwiches and a large pitcher of milk. Zevenbergen explained that Dura Vermeer was also working to construct buoyant roads and floating greenhouses. While each of these projects represents a somewhat different engineering challenge, they have a common goal, which is to allow people to continue to inhabit areas that, periodically at least, will be inundated. "There is a flood market emerging," Zevenbergen told me.

From the company's headquarters, it was about an hour's drive to Maasbommel. By the time I arrived, the sun was starting to sink, and in the afternoon light, the Meuse was glowing silver.

The amphibious homes all look alike. They are tall and narrow, with flat sides and curved metal roofs, so that standing next to one another they resemble a row of toasters. Each one is moored to a metal pole and sits on a set of hollow concrete pontoons. Assuming that all goes according to plan, when the Meuse floods, the homes will bob up and then, when the water recedes, they will gently be deposited back on land. At the point that I visited, a half a dozen families were occupying their amphibious houses. Anna van der Molen, a nurse and mother of four, gave me a tour of hers. She was enthusiastic about life on the river. "Not one day is the same," she told me. In the future, she said, she expected that people all over the world would live in floating houses, since, as she put it, "the water is coming up, and we have to live with it, not fight it—it's just not possible."

Chapter 7

BUSINESS AS USUAL

IN CLIMATE-SCIENCE CIRCLES, a future in which current emissions trends continue, unchecked, is known as "business as usual," or BAU. About five years ago, Robert Socolow, a professor of engineering at Princeton, began to think about BAU and what it implied for the fate of mankind. At that point, Socolow had recently become codirector of the Carbon Mitigation Initiative, a project funded by BP and Ford, but he still considered himself an outsider to the field of climate science. Talking to insiders, he was struck by the degree of their alarm. "I've been involved in a number of fields where there's a lay opinion and a scientific opinion," he told me when I went to visit him at his office shortly after returning from the Netherlands. "And, in most of the cases, it's the lay community that is more exercised, more anxious. If you take an extreme example, it would be nuclear power, where most of the people who work in nuclear science are relatively relaxed about very low levels of radiation. But, in the climate case, the experts—the people who work with the climate models every day, the people who do ice cores—they are *more* concerned. They're going out of their way to say, 'Wake up! This is not a good thing to be doing.'"

Socolow, who is sixty-seven, is a trim man with wire-rimmed glasses and gray, vaguely Einsteinian hair. Although by training he is a theoretical physicist—he did his doctoral research on quarks—he has spent most of his career working on problems of a more human scale, like how to prevent nuclear proliferation or construct buildings that don't leak heat. In the 1970s, Socolow helped design an energy-efficient housing development in Twin Rivers, New Jersey. At another point, he developed a system—never commercially viable—to provide air-conditioning in the summer using ice created in the winter. When Socolow became codirector of the Carbon Mitigation Initiative, he decided that the first thing he needed to do was get a handle on the scale of the carbon problem. He found that the existing literature on the subject offered almost too much information. In addition to BAU, a dozen or so alternative scenarios, known by code names like A1 and B1, had been devised; these all tended to jumble together in his mind, like so many Scrabble tiles. "I'm pretty quantitative, but I could not remember these graphs from one day to the next," he recalled. He decided to try to streamline the problem, mainly so that he could understand it.

Here in the United States, most of us begin generating CO_2 as soon as we get out of bed. Seventy percent of our electricity is generated by burning fossil fuels—a little more than 50 percent from burning coal and another 17 percent from natural gas—so that to turn on the lights is, indirectly at least, to pump carbon dioxide into the atmosphere. Making a pot of coffee, either on an electric or a gas range, adds more emissions, as does taking a hot shower, watching the morning news on TV, and driving to work. Exactly how much CO_2 any particular action produces depends on a variety of factors. Though all fossil fuels produce carbon dioxide as an inevitable product of combustion, some fuels, most notably coal, give off more than others for each unit of power generated. A kilowatt-hour of electricity delivered from a coal-fired plant will produce slightly more than half a pound of carbon, while if the power is originating from a plant that runs on natural gas, it will produce roughly half that amount. (When measuring CO_2, it is customary to count not the full weight of the gas, but just the weight of the carbon—to convert back, multiply by 3.7.) Every gallon of gasoline that is consumed produces about five pounds of carbon, meaning that in the course of a forty-mile commute, a vehicle like a Ford Explorer or a GM Yukon throws about a dozen pounds of carbon into the air. On average, every single person in America generates twelve thousand pounds of carbon per year. (If you would like to figure out your own annual contribution to greenhouse warming, go to the Environmental Protection Agency's Web site and plug various facts about your lifestyle—what kind of car you drive, how much of your trash you recycle, and so on—into the "personal emissions calculator" provided there.) The largest single source of carbon emissions in the United States is electricity production, at 39 percent, followed by transportation, at 32 percent. In a country like France, where three quarters of the power is produced by nuclear plants, this ratio is very different, and it's different again in countries like Bhutan, where many people don't even have access to electricity and where they burn wood and animal waste to cook and heat their homes.

In the future, the growth of carbon emissions is likely to be determined by several forces. One is the rate of population growth; estimates of how many people will be living on the planet in 2050 range from a low of 7.4 billion to a high of 10.6 billion. Another is economic growth. A third factor is the rate at which new technologies are adopted. Particularly in the developing world, the demand for electricity is increasing rapidly; in China, for example, electricity consumption is expected to more than double by 2025. If developing nations satisfy this demand by adopting the latest, most energy-efficient technologies, then emissions will grow at one rate. (This possibility is sometimes referred to as "leapfrogging," since it would require developing countries to "leapfrog" ahead of industrialized nations.) If they satisfy demand by deploying less efficient—but often cheaper—technologies, emissions will increase at a much faster rate.

"Business as usual" refers to a whole range of projections, all of which take as their primary assumption that emissions will continue to grow without regard to the climate. In 2005, global emissions amounted to roughly 7 billion metric tons of carbon. Under a midrange BAU projection, they will grow to 10.5 billion metric tons a year by 2029, and 14 billion tons a year by 2054. Under this same projection, CO_2 levels in the atmosphere will reach 500 parts per million by the middle of the century, and if things continue on the same trajectory, CO_2 will reach 750 parts per million, or roughly three times preindustrial levels, by the year 2100.

Looking at these figures, Socolow reached a couple of conclusions right away. The first was that to avoid exceeding CO_2 concentrations of 500 parts per million, immediate action would be needed. The second was that to meet this target, emissions growth would have to be held essentially to zero. Stabilizing CO_2 emissions would be such an enormous undertaking that Socolow decided to break the problem down into more manageable blocks, which he called "stabilization wedges." For simplicity's sake, he defined a stabilization wedge as a step that would be sufficient to prevent a billion metric tons of carbon per year from being emitted by 2054. Since annual carbon emissions now amount to 7 billion metric tons, and in fifty years are expected to reach 14 billion metric tons, seven wedges would be needed to hold emissions constant at today's level. With the help of a Princeton colleague, Stephen Pacala, Socolow eventually came up with fifteen different wedges— theoretically, at least, eight more than would be necessary. In August 2004, Socolow and Pacala published their findings in a paper in *Science* that received a great deal of attention. The paper was at once upbeat—"Humanity already possesses the fundamental scientific, technical, and industrial know-how to solve the carbon and climate problem for the next half-century," it declared—and deeply sobering. "There is no easy wedge" is how Socolow put it to me.

Consider wedge No. 11. This is the photovoltaic, or solar power, wedge—probably the most appealing of all the alternatives, at least in the abstract. Photovoltaic cells, which have been around for more than fifty years, are already in use in all sorts of small-scale applications and in some larger ones where the cost of connecting to the electrical grid is prohibitively high. The technology, once installed, is completely emissions-free, producing no waste products, not even water. For the purpose of their calculations, Socolow and Pacala assumed that a one-thousand-megawatt coal-fired power plant would produce about 1.5 million tons of carbon a year. (Today's coal plants actually produce some 2 million tons of carbon a year, but in the future, plants are expected to become more efficient.) To reduce emissions by a billion metric tons a year, enough solar cells would therefore have to be installed to obviate the need for nearly seven hundred thousand-megawatt coal plants. Since sunshine is not constant—it is interrupted by nightfall and by clouds—some two million megawatts of capacity would be needed. This, it

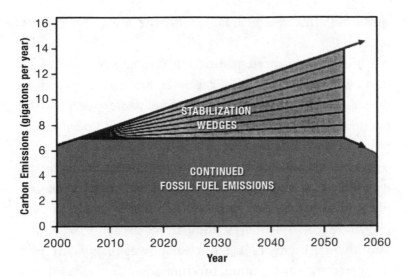

One "wedge" would prevent a billion tons of carbon a year from being emitted by 2054. Credit: S. Pacala and R. Socolow, Science, *vol. 305 (2004).*

turns out, would require PV arrays covering a surface area of five million acres—approximately the size of Connecticut.

Wedge No. 10 is wind electricity. Again, the technology has the advantage of being both safe and emissions-free. A large turbine can generate two megawatts of power, but since the wind, like sunlight, is intermittent, to get a wedge out of wind power would require at least a million two-megawatt turbines. Wind turbines are generally installed either offshore, or on hilltops or windy plains. When they are installed on land, the area around them can be used for other purposes, such as farming, but a million turbines would effectively "occupy" thirty million acres, an area roughly the size of New York state.

Other wedges present different challenges, some technical, some social. Nuclear power produces no carbon dioxide, but it generates radioactive waste, with all the attendant difficulties of storage, disposal, and international policing. More than forty years after the first commercial reactors went online, the United States has been unable to solve its nuclear waste problems, and several power plant operators have sued the federal government over its failure to construct a long-term waste storage site. Worldwide, there are 441 nuclear power plants currently in operation; one wedge could be achieved by doubling their capacity. There is also one heating and lighting wedge, which would result from cutting energy use in residential and commercial buildings by a quarter, and two automobile wedges. The first auto wedge would require that every car in the world be driven half as much as it is today, the second that it be twice as efficient. (Since the late 1980s, the fuel efficiency of passenger vehicles in the United States has actually declined, by more than 5 percent.)

Another possible option is a technology known as "carbon capture and storage," or CCS. As the name suggests, with CCS carbon dioxide is "captured" at the source—presumably a large emitter—and then injected at very high pressure into geological formations, such as depleted oil fields, underground. (At such pressure, CO_2 becomes "supercritical," a phase in which it is not quite a liquid and not quite a gas.) One wedge in Socolow's plan comes from "capturing" CO_2 from power plants, another from capturing it from synthetic-fuel manufacturers. The basic techniques of CCS are currently employed to increase production from oil and natural gas wells. However, at this point, there are no synthetic-fuel or power plants using the process. Nor does anyone know for certain how long CO_2 injected underground will remain there. The world's longest-running CCS effort, maintained by the Norwegian oil company Statoil at a natural gas field in the North Sea, has been operational only for about a decade. A wedge of CCS would require thirty-five hundred projects on the scale of Statoil's.

In a world like today's, where there is, for the most part, no direct cost to emitting CO_2, none of Socolow's wedges are apt to be implemented; this is, of course, why they represent a departure from "business as usual." To alter the economics against carbon requires government intervention. Countries could set a strict limit on CO_2, and then let emitters buy and sell carbon "credits." (In the United States, this same basic strategy has been used successfully with sulfur dioxide in order to curb acid rain.) Another alternative is to levy a tax on carbon. Both of these options have been extensively studied by economists; using their work, Socolow estimates that the cost of emitting carbon would have to rise to around a hundred dollars a ton to provide a sufficient incentive to adopt many of the options he has proposed. Assuming that the cost were passed on to consumers, a hundred dollars a ton would raise the price of a kilowatt-hour of coal-generated electricity by about two cents, which would add roughly fifteen dollars a month to the average American family's electricity bill.

All of Socolow's calculations are based on the notion—clearly hypothetical—that steps to stabilize emissions will be taken immediately, or at least within the next few years. This assumption is key not only because we are constantly pumping more CO_2 into the atmosphere but also because we are constantly building infrastructure that, in effect, guarantees that that much additional CO_2 will be released in the future. In the United States, the average new car gets about twenty miles to the gallon; if it is driven a hundred thousand miles, it will produce more than eleven metric tons of carbon. A thousand-megawatt coal plant built today, meanwhile, is likely to last fifty years and to emit some hundred million tons of carbon during its life. The overriding message of Socolow's wedges is that the longer we wait—and the more infrastructure we build without regard to its impact on emissions—the more daunting the task of keeping CO_2 levels below 500 parts per million will become.

Indeed, even if we were to hold emissions steady for the next half century, Socolow's graphs show that much steeper cuts would be needed in the following half century to keep CO_2 concentrations from exceeding that level. Carbon dioxide is a persistent gas; it lasts for about a century. Thus, while it is possible to increase CO_2 concentrations relatively quickly, the opposite is not the case. (The effect might be compared to driving a car equipped with an accelerator but no brakes.) After a while, I asked Socolow whether he thought that stabilizing emissions was a politically practical goal. He frowned.

"I'm always being asked, 'What can you say about the practicability of various targets?'" he told me. "I really think that's the wrong question. These things can all be done.

"What kind of issue is like this that we faced in the past?" he continued. "I think it's the kind of issue where something looked extremely difficult, and not worth it, and then people changed their minds. Take child labor. We decided we would not have child labor and goods would become more expensive. It's a changed preference system. Slavery also had some of those characteristics a hundred and fifty years ago. Some people thought it was wrong, and they made their arguments, and they didn't carry the day. And then something happened and all of a sudden it was wrong and we didn't do it anymore. And there were social costs to that. I suppose cotton was more expensive. We said, 'That's the trade-off; we don't want to do this anymore.' So we may look at this and say, 'We are tampering with the earth.' The earth is a twitchy system. It's clear from the record that it does things that we don't fully understand. And we're not going to understand them in the time period we have to make these decisions. We just know they're there. We may say, 'We just don't want to do this to ourselves.' If it's a problem like that, then asking whether it's practical or not is really not going to help very much. Whether it's practical depends on how much we give a damn."

Marty Hoffert is a professor of physics at New York University. He is big and bearish, with a wide face and silvery hair. Hoffert got his undergraduate degree in aeronautical engineering, and one of his first jobs, in the mid-1960s, was helping to develop the United States's antiballistic-missile system. During the week, Hoffert worked at a lab in New York, and sometimes he would go down to Washington to meet with Pentagon officials. Over the weekend, on occasion, he would travel back to Washington to protest Pentagon policy. Eventually, he decided that he wanted to work on something, in his words, "more productive." In this way, he became involved in climate research. He calls himself a "technological optimist," and a lot of his ideas about electric power have a wouldn't-it-be-cool, Buck Rogers sound to them. On other topics, though, Hoffert is a killjoy.

"We have to face the quantitative nature of the challenge," he told me one day over lunch at the NYU faculty club. "Right now, we're going to just burn everything up; we're going to heat the atmosphere to the temperature it was in the Cretaceous, when there were crocodiles at the poles. And then everything will collapse."

Hoffert is primarily interested in finding new, carbon-free ways to generate energy. Currently, the technology that he is pushing is space-based solar power, or SSP. In theory, at least, SSP involves launching into space satellites equipped with massive photovoltaic arrays. Once a satellite is in orbit, the array would unfold or, according to some plans, inflate. SSP has two important advantages over conventional, land-based solar power. In the first place, there is more sunlight in space—roughly eight times as much, per unit of area—and, in the second, this sunlight is constant: satellites are not affected by clouds or by nightfall. The obstacles, meanwhile, are several. No full-scale test of SSP has ever been conducted. (In the 1970s, NASA studied the idea of sending a photovoltaic array the size of Manhattan into space, but the project never, as it were, got off the ground.) Then, there is the expense of launching satellites. Finally, once the arrays are up, there is the difficulty of getting the energy down. Hoffert imagines solving this last problem by using microwave beams of the sort used by cell phone towers, only much more tightly focused. He believes, as he put it to me, that SSP has a great deal of "long-term promise"; however, he is quick to point out that he is open to other ideas, like putting solar collectors on the moon, or using superconducting wires to transmit electricity with minimal energy loss, or generating wind power using turbines suspended in the jet stream. The important thing, he says, is not *which* new technology will work but simply that *some* new technology be found: "There's an argument that our civilization can continue to exist with the present number of people and the present kind of high technology through conservation. I see that argument as similar to a man being locked in a sealed room with a limited amount of oxygen. And if he breathes more slowly, he'll be able to live longer, but what he really needs is to get out of the room. And I want to get out of the room." A few years ago, Hoffert published an influential paper in *Science* in which he argued that holding CO_2 levels below 500 parts per million would require a "Herculean" effort and probably could be accomplished only through "revolutionary" changes in energy production.

"The idea that we already possess the 'scientific, technical, and industrial know-how to solve the carbon problem' is true in the sense that, in 1939, the technical and scientific expertise to build nuclear weapons existed," he told me, quoting Socolow. "But it took the Manhattan Project to make it so."

Hoffert's primary disagreement with Socolow, which both men took pains to point out to me and also took pains to try to minimize, is over the future trajectory of CO_2 emissions. For the past several decades, as the world has turned increasingly from coal to oil, natural gas, and nuclear power, emissions of CO_2 per unit of energy have declined, a process known as "decarbonization." This has slowed the growth of emissions relative to the growth of the global economy; without it, CO_2 levels today would be significantly higher.

In the "business as usual" scenario that Socolow uses, it is assumed that decarbonization will continue. To assume this, however, is to overlook several emerging trends. Most of the growth in energy usage in the next few decades is due to occur in places like China and India, where supplies of coal far exceed those of oil or natural gas. (China, which is adding new coal-fired generating capacity at the rate of more than a gigawatt a month, is expected to overtake the United States as the world's largest carbon emitter around 2025.) Meanwhile, global production of oil and gas is expected to start to decline—according to some experts in twenty or thirty years, and to others by the end of this decade. Hoffert predicts that the world will start to "recarbonize," a development that would make the task of stabilizing carbon dioxide that much more difficult. By his accounting, recarbonization will mean that as many as twelve wedges will be needed simply to keep CO_2 emissions on the same upward trajectory they're on now. (Socolow readily acknowledges that there are plausible scenarios that would push up the number of wedges needed.) Hoffert told me that he thought the federal government should be budgeting between ten and twenty billion dollars a year for primary research into new energy sources. For comparison's sake, he pointed out that the "Star Wars" missile-defense program, which still hasn't yielded a workable system, has already cost the government nearly a hundred billion dollars.

A commonly heard argument against acting to curb

global warming is that the options now available are inadequate. To his dismay, Hoffert often finds his ideas being cited in support of this argument, with which, he says, he vigorously disagrees. "I want to make it very clear," he told me at one point. "We have to start working immediately to implement those elements that we know how to implement *and* we need to start implementing these longer-term programs. Those are not opposing ideas."

"Let me say this," he said at another point. "I'm not sure we can solve the problem. I hope we can. I think we have a shot. I mean, it may be that we're not going to solve global warming, the earth is going to become an ecological disaster, and, you know, somebody will visit in a few hundred million years and find there were some intelligent beings who lived here for a while, but they just couldn't handle the transition from being hunter-gatherers to high technology. It's certainly possible. Carl Sagan had an equation—the Drake equation—for how many intelligent species there are in the galaxy. He figured it out by saying, How many stars are there, how many planets are there around these stars, what's the probability that life will evolve on a planet, what's the probability if you have life evolve of having intelligent species evolve, and, once that happens, what's the average lifetime of a technological civilization? And that last one is the most sensitive number. If the average lifetime is about a hundred years, then probably, in the whole galaxy of four hundred billion stars, there are only a few that have intelligent civilizations.

If the lifetime is several million years, then the galaxy is teeming with intelligent life. It's sort of interesting to look at it that way. And we don't know. We could go either way."

Chapter 8

THE DAY AFTER KYOTO

W HEN THE KYOTO PROTOCOL went into effect, on February 16, 2005, the event was seen as a cause for celebration in many cities around the world. The mayor of Bonn hosted a reception in the Rathaus; Oxford University held an "Entry into Force" banquet; and in Hong Kong there was a Kyoto prayer meeting. As it happened, that day, an exceptionally warm one in Washington, D.C., I went to speak to the Under Secretary of State for Democracy and Global Affairs, Paula Dobriansky.

Dobriansky is a slight woman with shoulder-length brown hair and a vaguely anxious manner. Among her duties is explaining the Bush administration's position on global warming to the rest of the world, a task that, on the occasion of Kyoto's entry into force, seemed peculiarly unenviable. The United States is by far the world's largest emitter of greenhouse gases in aggregate—it produces nearly a quarter of the world's total—and on a per capita basis is rivaled only by a handful of nations, like Qatar. Yet the United States is one of only two industrialized nations that have rejected the Kyoto Protocol, and, with it, mandatory cuts in emissions. (The other outlier is Australia.) Two of Dobriansky's assistants accompanied me into her office. We all took seats in a circle.

Dobriansky began by assuring me that despite how it might appear, the Bush administration took the issue of climate change "very seriously." She went on, "Also let me just add, because in terms of taking it seriously, not only stating to you that we take it seriously, we have engaged many countries in initiatives and efforts, whether they are bilateral initiatives—we have some fourteen bilateral initiatives—and in addition we have put together some multilateral initiatives. So we view this as a serious issue." I asked her how, then, the administration justified its position on Kyoto to its allies. "We have a common goal and objective," she replied. "Where we differ is on what approach we believe is and can be the most effective." A few moments later, she added, as if expanding on this statement: "The bottom line here is, in grappling with a serious issue, we believe we have a common goal and objective, but that we can take different approaches."

The remainder of our brief conversation followed much the same lines. At one point, I asked the undersecretary if there were any circumstances under which the administration would accede to mandatory caps on emissions. "Our approach has been predicated on: we act, we learn, we act again," she said. In response to a question about how urgent the problem of stabilizing emissions was, she replied, "We act, we learn, we act again," and in response to a question about what would constitute a "dangerous" level of CO_2 in the atmosphere, she said, "Forgive me, I'm going to repeat myself: we act, we learn, we act again." Dobriansky told me twice that the administration's approach to global warming encompassed both "near-term actions and long-term actions" and three times that it saw economic growth as "the solution, not the problem." I had been instructed that Dobriansky could spare no more than twenty minutes. According to my digital recorder, after fifteen minutes and thirty-five seconds one of her assistants announced that it was time to wrap things up. As I was getting ready to leave, I asked Dobriansky if there was anything more she wanted to say.

"I'd say this to you," she replied. "We see this as a serious issue. We have vigorously and robustly put forth a climate change policy to address these issues, and we will continue to work with other countries to address the issue of climate change. Basically and fundamentally we have a common goal and objective, but we are pursuing different approaches."

On paper at least, the United States, along with the rest of the world, has been committed to addressing global warming for nearly fifteen years. In June 1992, the United Nations held the so-called Earth Summit in Rio de Janeiro, which was attended by more than twenty thousand people. Representatives from virtually every country on the globe met there to discuss and ultimately endorse the U.N. Framework Convention on Climate Change. One of the earliest signatories was President George H. W.

Bush, who, while in Rio, called on world leaders to translate "the words spoken here into concrete action to protect the planet." Three months later, Bush submitted the Framework Convention to the U.S. Senate, which approved it by unanimous consent.

In the English version, the Framework Convention runs to thirty-three pages. It starts with vague statements of principle ("Acknowledging that change in the Earth's climate and its adverse effects are a common concern of humankind . . ."; "Concerned that human activities have been substantially increasing the atmospheric concentrations of greenhouse gases . . .") and works its way through a long list of definitions (" 'Climate change' means a change of climate which is attributed directly or indirectly to human activity"; " 'Climate system' means the totality of the atmosphere, hydrosphere, biosphere and geosphere and their interactions") before finally arriving at its objective. This is: the "stabilization of greenhouse gas concentrations in the atmosphere at a level that would prevent dangerous anthropogenic interference with the climate system."

Every country that signed on to the Framework Convention accepted the same goal—avoiding DAI. But not every country accepted the same obligations. The treaty distinguished between industrialized nations, which, in U.N.-speak, became known as the Annex 1 countries, and basically everyone else. While the latter group agreed to take steps to "mitigate" climate change, the former agreed to reduce its greenhouse gas emissions. (In diplomatic terms, this arrangement followed the principle of "common but differentiated responsibilities.") Article 4, paragraph 2, subparagraph b of the Framework Convention spelled out what compliance meant; it instructed Annex 1 countries, which include the United States, Canada, Japan, and the nations of Europe and the erstwhile Eastern bloc—to "aim" to return their emissions to 1990 levels.

As it turned out, submitting the Framework Convention to the Senate was one of George Bush Senior's last acts as president. Bill Clinton reaffirmed U.S. support of the convention, announcing, shortly after taking office, on Earth Day 1993, that the nation was committed to reducing its greenhouse gas emissions to 1990 levels by the year 2000. "Unless we act now," he said, "we face a future in which the sun may scorch us, not warm us; where the change of season may take on a dreadful new meaning; and where our children's children will inherit a planet far less hospitable than the world in which we came of age."

Yet even as Clinton was reasserting the nation's commitment, emissions in the United States and indeed around the globe were continuing to rise. By 1995, pretty much the only countries that were making any progress toward compliance were former members of the Soviet bloc, and this was because their economies were in free fall. Meanwhile, as emissions continued to go up, what had initially seemed a rather modest goal—returning to 1990 levels— started to look more and more ambitious. Several rounds of often bitter negotiations followed—in Berlin in March

1995, in Geneva in July 1996, and, finally, in Kyoto in December 1997.

Technically, the agreement that emerged from the Kyoto session is simply an addendum to the Framework Convention. (Its full title is the Kyoto Protocol to the United Nations Framework Convention on Climate Change.) The protocol has the same goal as the convention—avoiding DAI—and hews to the same principle of "common but differentiated responsibilities." But for vague exhortations, like "aim," the protocol substitutes mandatory commitments. Exactly what these commitments are varies slightly from country to country, based on a combination of historical and political factors. The nations of the European Union, for example, are supposed to reduce their greenhouse gas emissions 8 percent below 1990 levels and to do so by 2012, the year that the protocol lapses. The United States, meanwhile, has a target of 7 percent below 1990 levels, and Japan has a target of 6 percent below. The treaty covers five greenhouse gases in addition to CO_2—methane, nitrous oxide, hydrofluorocarbons, perfluorocarbons, and sulfur hexafluoride—which, for the purposes of accounting, are converted into units known as "carbon dioxide equivalents." Annex 1 nations can meet their targets, in part, by buying and selling emissions "credits" and by investing in "clean development" projects in non–Annex 1 nations, like China and India.

Even as Kyoto was being negotiated, it was clear that the treaty was going to face opposition from many of the same senators who had voted in favor of the original Framework Convention. In July of 1997, Senator Chuck Hagel, Republican of Nebraska, and Senator Robert Byrd, Democrat of West Virginia, introduced a "sense of the Senate" resolution publicly warning the Clinton administration against the direction that the talks were taking. The so-called Byrd-Hagel Resolution stated that the United States should reject any agreement that committed it to reducing emissions unless concomitant obligations were imposed on developing countries as well. The Senate approved the resolution by a vote of 95–0, an outcome that reflected lobbying by both business and labor. (The Global Climate Coalition, a group that was sponsored by, among others, Chevron, Chrysler, Exxon, Ford, General Motors, Mobil, Shell, and Texaco, spent some $13 million on an anti–Kyoto Protocol advertising campaign.)

From a certain perspective, the logic behind the Byrd-Hagel Resolution is unimpeachable. Emissions controls cost money, and this cost has to be borne by somebody. If the United States were to agree to limit its greenhouse gases while economic competitors like China and India did not, then American companies would be put at a disadvantage. "A treaty that requires binding commitments for reduction of emissions of greenhouse gases for the industrial countries but not developing countries will create a very damaging situation for the American economy" is how Richard Trumka, the secretary-treasurer of the AFL–CIO, put it when he traveled to Kyoto to lobby against the protocol. It is also true that an agreement that limits carbon emissions in some countries and not in others could result in a migration, rather than an actual reduction, of CO_2 emissions. (Such a possibility is known in climate parlance as "leakage.")

From another perspective, however, the logic of Byrd-Hagel is deeply, even obscenely, self-serving. Suppose for a moment that the total anthropogenic CO_2 that can be emitted into the atmosphere were a big ice-cream cake. If the aim is to keep global concentrations below five hundred parts per million, then roughly half that cake has already been consumed, and, of that half, the lion's share has been polished off by the industrialized world. To insist now that all countries cut their emissions simultaneously amounts to advocating that industrialized nations be allocated most of the remaining slices, on the ground that they've already gobbled up so much. In a year, the average American produces the same greenhouse-gas emissions as four and a half Mexicans, or eighteen Indians, or ninety-nine Bangladeshis. If both the United States and India were to reduce their emissions proportionately, then the average Bostonian could continue indefinitely producing eighteen times as much greenhouse gases as the average Bangalorean. But why should anyone have the right to emit more than anyone else? At a climate meeting in New Delhi a few years ago, Atal Bihari Vajpayee, then the Indian prime minister, told world leaders, "Our per capita greenhouse gas emissions are only a fraction of the world average and an order of magnitude below that of many developed countries. We do not believe that the ethos of democracy can support any norm other than equal per capita rights to global environmental resources."

Outside the United States, the decision to exempt developing nations from Kyoto's mandates was generally regarded as an adequate—if imperfect—solution to an otherwise intractable problem. The arrangement was basic to the Framework Convention, and it mimicked a structure that had already been employed—successfully—to deal with another potential global crisis: the depletion of atmospheric ozone. The Montreal Protocol, adopted in 1987, called for a phase-out of ozone-depleting chemicals, but gave developing nations what amounted to a ten-year grace period. Pieter van Geel, the Dutch environment secretary, described the European outlook to me as follows: "We cannot say, 'Well, we have our wealth, based on the use of fossil fuels for the last three hundred years, and, now that your countries are growing, you may not grow at this rate, because we have a climate-change problem.' We should show moral leadership by giving the example. That's the only way we can ask something of these other countries."

For its part, the Clinton administration supported the Kyoto Protocol in theory, but not really in practice. In November 1998, the United States's ambassador to the U.N. signed the treaty on behalf of the administration. But the president never submitted it to the Senate, where clearly it wouldn't have won the two-thirds vote needed for ratification. On Earth Day 2000, Clinton delivered more or less the same speech he had given seven years earlier: "The greatest environmental challenge of the new century is global warming. The scientists tell us the 1990s were the hottest decade of the entire millennium. If we fail to reduce the emission of greenhouse gases, deadly heat waves and droughts will become more frequent, coastal areas will flood, and economies will be disrupted. That is going to happen, unless we act." By the time he left office, CO_2 emissions from the United States were 15 percent higher than they had been in 1990.

No politician in America—perhaps no major politician in the world—is more closely associated with the subject of global warming than Al Gore. In 1992, while still in the Senate, Gore published *Earth in the Balance*, in which he argued that protecting the global environment should be the "central organizing principle" of society; five years later, as vice president, he flew to Japan to salvage Kyoto when negotiations seemed on the verge of breaking down. Nonetheless, global warming never really became a factor in the 2000 election. During the campaign, George W. Bush repeatedly asserted that he, too, was deeply concerned about climate change, calling it "an issue that we need to take very seriously." He promised that, if elected, he would impose federal regulations limiting CO_2 emissions.

Soon after his inauguration, Bush sent the new head of the Environmental Protection Agency, Christine Todd Whitman, to a meeting of environmental ministers from the world's leading industrialized nations, where she elaborated on what she apparently believed to be his position. Whitman assured her colleagues that the president considered global warming to be "one of the greatest environmental challenges that we face" and that he wanted to "take steps to move forward." Ten days after her presentation, Bush announced that not only was he withdrawing the United States from the ongoing negotiations over Kyoto—the protocol had left several complex issues of implementation to be resolved later—but also he had changed his mind about federal curbs on carbon dioxide. Explaining this reversal, Bush asserted that he no longer thought CO_2 limits were justified, owing to the "state of scientific knowledge of the causes of, and solutions to, global climate change," which he labeled "incomplete." (Former Treasury Secretary Paul O'Neill, who backed the president's original position, has speculated publicly that the reversal was engineered by Vice President Dick Cheney.)

For nearly a year, the Bush administration operated essentially without any position on climate change. Then, the president announced that the United States would be pursuing a whole new approach. Instead of focusing on greenhouse gas emissions, the country would focus on something called "greenhouse gas intensity." Bush declared this new approach preferable because it recognized "that a nation that grows its economy is a nation that can afford investments and new technology."

Greenhouse gas intensity is not a quantity that can be measured directly. Rather, it is a ratio that relates emissions to economic output. Say, for example, that one year a business produces a hundred pounds of carbon and a hundred dollars' worth of goods. Its greenhouse gas intensity in that case would be one pound per dollar. If the next year that company produces the same amount of carbon but an extra dollar's worth of goods, its intensity will have fallen by one percent. Even if it doubles its total emissions of carbon, a company—or a country—can still claim a reduced intensity provided that it more than doubles its output of goods. (Typically, a country's greenhouse gas intensity is measured in terms of tons of carbon per million dollars' worth of gross domestic product.)

To focus on greenhouse gas intensity is to give a peculiarly sunny account of the U.S. situation. Between 1990 and 2000, U.S. greenhouse gas intensity fell by some 17 percent, owing to several factors, including the shift toward a more service-based economy. Meanwhile, total emissions rose by some 12 percent. (In terms of greenhouse gas intensity, the United States actually performs better than many third world nations, because even though we consume a lot more energy, we also have a much larger GDP.) In February 2002, President Bush set the goal of reducing the country's greenhouse gas intensity by 18 percent over the following ten years. During that same decade, his administration expects the American economy to grow by 3 percent annually. If both expectations are met, overall emissions of greenhouse gases will rise by about 12 percent.

The administration's plan, which relies almost entirely

on voluntary measures, has been characterized by critics as nothing more than a subterfuge—"a total charade" is how Philip Clapp, president of the Washington-based National Environmental Trust, once put it. And certainly, if the goal is to prevent "dangerous anthropogenic interference," then greenhouse gas intensity is the wrong measure to use. (Essentially, the president's approach amounts to following the path of "business as usual.") The administration's response to such criticism has generally been to attack its premise. "Science tells us that we cannot say with any certainty what constitutes a dangerous level of warming and therefore what level must be avoided," Paula Dobriansky has stated. When I asked her how, in that case, the United States could support the aim of averting DAI, she answered by saying—twice—"We predicate our policies on sound science."

Right around the time I went to visit Dobriansky, the chairman of the Senate Committee on Environment and Public Works, James Inhofe, gave a speech on the Senate floor, which he titled "An Update on the Science of Climate Change." In the speech, Inhofe, an Oklahoma Republican, announced that "new evidence" had come to light that "makes a mockery" of the notion that human-induced warming is occurring. The senator, who has called global warming "the greatest hoax ever perpetrated on the American people," went on to argue that this important new evidence was being suppressed by "alarmists" who view anthropogenic warming as "an article of religious faith." One of the authorities that Inhofe repeatedly cited in support of his claims was the fiction writer Michael Crichton.

It was an American scientist—Charles David Keeling— who, in the 1950s, developed the technology to measure CO_2 levels precisely, and it was American researchers who, working on Mauna Loa, first showed that these levels were steadily rising. In the half century since then, the United States has contributed more than any other nation to the advancement of climate science, both theoretically, through the work of climate modelers at places like GISS and NOAA's Geophysical Fluid Dynamics Laboratory, and experimentally, through field studies conducted in the Arctic, the Antarctic, and every continent in between.

At the same time, the United States is also the world's chief purveyor of the work of so-called global-warming skeptics. The ideas of these skeptics are published in books with titles like *The Satanic Gases* and *Global Warming and Other Eco-Myths* and then circulated on the Web by groups like Tech Central Station, which is sponsored by, among others, ExxonMobil and General Motors. While some skeptics' organizations argue that global warming isn't real, or at least hasn't been proved—"Predicting *weather* conditions a day or two in advance is hard enough, so just imagine how hard it is to forecast what our *climate* will be," Americans for Balanced Energy Choices, a lobbying organization funded by mining and power companies, declares on its Web site—others maintain that rising CO_2 levels are actually cause for celebration.

"Carbon dioxide emissions from fossil fuel combustion are beneficial to life on earth," the Greening Earth Society, an organization created by the Western Fuels Association, a utility group, states. Atmospheric levels of 750 parts per million—nearly triple preindustrial levels—are nothing to worry about, the society maintains, because plants like lots of CO_2, which they need for photosynthesis. (Research on this topic, the group's Web site acknowledges, has been "frequently denigrated," but "it's exciting stuff" and provides an "antidote to gloom-and-doom about potential changes in earth's climate.")

In legitimate scientific circles, it is virtually impossible to find evidence of disagreement over the fundamentals of global warming. Naomi Oreskes, a professor of history and science studies at the University of California at San Diego, recently tried to quantify the level of consensus. She conducted a study of more than nine hundred articles on climate change published in refereed journals between 1993 and 2003 and subsequently made available on a leading research database. Of these, she found that 75 percent endorsed the view that anthropogenic emissions were responsible for at least some of the observed warming of the past fifty years. The remaining 25 percent, which dealt with questions of methodology or climate history, took no position on current conditions. Not a single article disputed the premise that anthropogenic warming is under way.

Still, pronouncements by groups like the Greening Earth Society and politicians like Senator Inhofe shape the public discourse on climate change. And this clearly is their point. A few years ago, pollster Frank Luntz prepared a strategy memo for Republican members of Congress, coaching them on how to deal with a variety of environmental issues. (Luntz, who first made a name for himself by helping to craft Newt Gingrich's "Contract with America," has been described as "a political consultant viewed by Republicans as King Arthur viewed Merlin.") Under the heading "Winning the Global Warming Debate," Luntz wrote, "The scientific debate is closing (against us) but not yet closed. There is still a window of opportunity to challenge the science." He warned, "Voters believe that there is *no consensus* about global warming in the scientific community. Should the public come to believe that the scientific issues are settled, their views about global warming will change accordingly." Luntz also advised, "The most important principle in any discussion of global warming is your commitment to sound science."

It is in this context, and really only in this context, that the Bush administration's claims about the science of global warming make any sense. Administration officials are quick to point to the scientific uncertainties that remain about global warming, of which there are many. But where there is broad agreement, they are reluctant to acknowledge it.

"When we make decisions, we want to make sure we do so on sound science," the president said, announcing his new approach to global warming in February 2002. Just a few months later, the Environmental Protection Agency delivered a two hundred and sixty-three page report to the U.N. that stated, "Continuing growth in greenhouse gas emissions is likely to lead to annual average warming over the United States that could be as much as several degrees Celsius (roughly 3 to 9 degrees Fahrenheit) during the 21st century." The president dismissed the report—the product of years of work by federal researchers—as something "put out by the bureaucracy." The following spring, the EPA made another effort to give an objective summary of climate science, in a report on the state of the environment. The White House interfered so insistently in the writing of the global warming section—at one point, it tried to insert excerpts from a study partly financed by the American Petroleum Institute—that, in an internal memo, agency staff members complained that the section "no longer accurately represents scientific consensus." (When the EPA finally published the report, the climate-science section was missing entirely.) In June 2005, the *New York Times* revealed that a White House official named Philip Cooney had repeatedly edited government reports on climate change in order to make their findings seem less alarming. In one instance, Cooney received a report stating: "Many scientific observations point to the conclusion that the Earth is undergoing a period of relatively rapid change." He revised this statement to read: "Many scientific observations <u>indicate</u> that the Earth <u>may be</u> undergoing a period of relatively rapid change." Shortly after his editing efforts were disclosed, Cooney resigned from his White House post and took a job with Exxon-Mobil.

On the day after the Kyoto Protocol took effect, the United Nations hosted a conference titled, appositely, "One Day After Kyoto." The conference, whose subtitle was "Next Steps on Climate," was held in a large room with banks of curved desks, each equipped with a little plastic earpiece. The speakers included scientists, insurance-industry executives, and diplomats from all over the world, among them the U.N. ambassador from the tiny Pacific island nation of Tuvalu, who described how his country was in danger of simply disappearing. Britain's permanent representative to the U.N., Sir Emyr Jones Parry, began his remarks to the crowd of two hundred or so by stating, "We can't go on as we are."

When the United States withdrew from negotiations over Kyoto, in 2001, the entire effort nearly collapsed. All on its own, America accounts for 34 percent of Annex 1 emissions. According to Kyoto's elaborate ratification mechanism, in order to take effect the protocol had to be approved by countries responsible for at least 55 percent of those emissions. European leaders spent more than three years working behind the scenes, trying to line up support from the remainder of the industrialized world. The crucial threshold was finally crossed in October 2004, when the Russian Duma voted in favor of ratification. The Duma's vote was all but explicitly understood to be a condition of European backing for Russia's bid to join the World

Trade Organization. (RUSSIA FORCED TO RATIFY KYOTO
PROTOCOL TO BECOME WTO MEMBER, read the headline in
Pravda.)

As speaker after speaker at the U.N. conference noted,
Kyoto is an important first step, but only a first step. The
protocol expires in 2012 and the cuts it mandates don't
come close to stabilizing worldwide emissions. Even if
every country—including the United States—were to
fulfill its obligations under Kyoto, CO_2 concentrations
in the atmosphere would still be headed to five hundred
parts per million, and beyond. Without substantive commitments from countries like China and India, there is no
realistic way to avoid DAI. But why should China and
India accept the costs of controlling emissions when
America has refused to do so? In this way, the United
States, having failed to defeat Kyoto, may be in the process
of doing something even more damaging: ruining the
chances of reaching a post-Kyoto agreement. "The blunt
reality is that, unless America comes back into some form
of international consensus, it is very hard to make progress"
is how Britain's prime minister, Tony Blair, recently put it.

Astonishingly, standing in the way of this progress seems
to be Bush's goal. Dobriansky explained the administration's position to me as follows: While the rest of the
industrialized world is pursuing one strategy (emissions
limits), the United States is pursuing another (no emissions
limits), and it is still too early to say which approach will
work best. "It is essential to really implement these
programs and approaches now and to take stock of their

effectiveness," she said, adding, "we think it is premature to talk about future arrangements." At a round of international climate talks held in Buenos Aires in December 2004, many delegations were pressing for a preliminary round of meetings so that work could start on mapping out a successor to the Kyoto Protocol. The U.S. delegation opposed these efforts so adamantly that finally the Americans were asked to describe, in writing, what sort of meeting they would find acceptable. They issued half a page of conditions, one of which was that the session "shall be a one-time event held during a single day." Another condition was, paradoxically, that, if they were going to discuss the future, the future would have to be barred from discussion; presentations, they wrote, should be limited to "an information exchange" on "existing national policies." Annie Petsonk, a lawyer with the advocacy group Environmental Defense, who previously worked for the administration of George Bush Senior, attended the talks in Buenos Aires. She recalled the effect that the memo had on the members of the other delegations: "They were ashen."

European leaders have made no secret of their dismay at the administration's stance. "It's absolutely obvious that global warming has started," France's president, Jacques Chirac, said after attending the 2004 summit of leaders of the world's major industrial powers—the Group of 8. "And so we have to act responsibly, and, if we do nothing, we would bear a heavy responsibility. I had the chance to talk to the United States president about this. To tell you that I convinced him would be a total exaggeration, as you can imagine." Tony Blair, who held the presidency of the G8 in 2005, spent the months leading up to that year's summit trying to convince Bush that, in his words, "the time to act is now." It's plain, Blair said in an address devoted to climate change, that "the emission of greenhouse gases . . . is causing global warming at a rate that began as significant, has become alarming, and is simply unsustainable in the long-term. And by 'long-term' I do not mean centuries ahead. I mean within the lifetime of my children certainly; and possibly within my own. And by 'unsustainable,' I do not mean a phenomenon causing problems of adjustment. I mean a challenge so far-reaching in its impact and irreversible in its destructive power, that it alters radically human existence." Just a few weeks before the 2005 summit, which was held in Gleneagles, Scotland, the national science academies of all the G8 nations, including the United States, along with the science academies of China, India, and Brazil, issued a remarkable joint statement calling on world leaders to "acknowledge that the threat of climate change is clear and increasing."

All of this, however, had no apparent impact on the president. In the lead-up to the summit, the head of the White House Council on Environmental Quality, James Connaughton, attended a meeting in London where he announced that he still wasn't convinced that anthropogenic warming was a problem. "We are still working on the issue of causation, the extent to which humans are a factor," he said. According to the *Washington Post*, administration officials insisted on weakening a proposal for joint action prepared for the summit, demanding, for example, the deletion of a passage citing "increasingly compelling evidence of climate change, including rising ocean and atmospheric temperatures, retreating ice sheets and glaciers, rising sea levels, and changes to ecosystems." The final communiqué from the summit, which was overshadowed by the London subway bombings, largely reflected the administration's position; it labeled global warming a "serious and long-term challenge" but also cited "uncertainties" in "our understanding of climate science" and called vaguely on G8 members to "promote innovation" and "accelerate deployment of cleaner technologies."

Senator John McCain, Republican of Arizona, is the primary sponsor of a bill that would, in effect, make good on George Bush's unfulfilled 2000 campaign promise to regulate carbon emissions. The Climate Stewardship Act calls for a reduction of greenhouse gas emissions in the United States to 2000 levels by 2010, and to 1990 levels by 2016. McCain has managed to get the Climate Stewardship Act onto the Senate floor for a vote twice, both times over strong White House opposition. In October 2003, the measure was defeated by a vote of fifty-five to forty-three; in June 2005, it went down sixty to thirty-eight. When I asked McCain to characterize Bush's position on global warming, he responded, "MIA."

"This is clearly an issue that we will win on over time because of the evidence," he went on. "The overwhelming impacts of climate change are becoming more and more visible every day. The problem is: will it be too late? We are a country that emits nearly 25 percent of the world's greenhouse gases. How much damage will have been done before we act?"

As of this writing, U.S. emissions are nearly 20 percent higher than they were in 1990.

Chapter 9

BURLINGTON, VERMONT

B URLINGTON, VERMONT, on the eastern shore of Lake Champlain, is by almost any measure a small city; still, it is the largest in Vermont. Several years ago, its voters decided that instead of authorizing the local utility company to buy more power, they would use less of it. Since then, the city has probably done as much as any municipality in the country to try to reduce its greenhouse gas emissions. The Burlington Electric Department may be the only utility in the United States whose vehicle fleet includes mountain bikes.

Peter Clavelle has been Burlington's mayor since 1989, with a two-year hiatus, which he likes to refer to as a "voter-inspired sabbatical." He is short and bald, with a salt-and-pepper mustache and mournful blue eyes. During his "sabbatical," Clavelle went to live with his family on the island of Grenada.

"Living on an island, you really get in touch with practices that are sustainable and practices that are unsustainable," he told me. It was a sticky July day, and we were driving around town in Clavelle's hybrid Honda Civic, looking at energy-saving projects. He paused to point out a city bus equipped with a bicycle rack on the front grille.

"The issues around climate protection are about sustainability," he went on. "They're about future generations. They're also about this conviction that local action does make a difference. Many of us are very frustrated with the lack of vision and action by the federal government, but there's a choice to be made. You either can bemoan federal policies or you can take control of your own destiny."

Burlington's energy-saving campaign, launched in 2002, is known as the "10 percent challenge." ("Put the chill on global warming" is its slogan.) As the name of the campaign suggests, the city's aim is to reduce greenhouse gas emissions by 10 percent, though from what baseline is somewhat vaguely defined. To further this goal, Burlington has tried just about everything, from providing free energy consultations to businesses to designing "energy efficiency calendars" for kids. Tray liners printed up for the local McDonald's feature a well-meaning but creepy-looking dinosaur named Climo Dino. "While our climate was changed by a giant asteroid, you humans are changing your own climate by emitting six billion tons of CO_2 into the atmosphere each year," Climo Dino observes.

The first stop on Clavelle's tour was an outpost of the county dump where, instead of collecting rubbish, the city sells it. Burlington encourages contractors to engage not in demolition but in "deconstruction," a practice that saves energy both by reducing the city's waste stream and cutting down on the need for new materials. Dozens of "deconstructed" sinks and doors and vanities were arrayed, showroom style, in what once had been a garage. A next-to-new staircase was leaning against the wall, waiting for a buyer needing steps of precisely the same dimensions. In the parking lot, some kids were building a garden shed out of old plywood. Clavelle told me that he had gotten the idea for ReCycle North from a similar program in Minneapolis. "It's management by plagiarism," he announced cheerfully.

Our next stop was the headquarters of the Burlington Electric Department, or BED for short. Behind the building, I could see a single wind turbine, which was turning briskly in the breeze. The turbine symbolizes the city's effort, while at the same time generating enough power for thirty homes. All in all, Burlington Electric gets nearly half of its energy from renewable sources, including a fifty-megawatt power plant in town that runs off wood chips. As we headed inside the BED headquarters, we passed a display of compact fluorescent light bulbs, which the company leases to interested customers at a cost of twenty cents a month. An electric department official named Chris Burns came out to greet us. He explained that a family that was keeping a hundred-watt incandescent bulb burning out on the porch all night could cut its electricity bill by up to 10 percent by simply replacing that bulb with a compact fluorescent. He said that several businesses in Burlington had cut their energy usage by significantly more than that just by taking such basic steps as adjusting the thermostat. The Burlington Electric Department has estimated that the energy-saving projects that the city has undertaken will, over the course of their useful life, prevent the release of nearly 175,000 tons of carbon. "We consider every building a power plant," Burns told me.

A little later in the day, Clavelle took me to visit the City Market, a grocery store built on municipal land that had previously been a hazardous waste site. The city supports the market so that Burlington residents won't have to drive to the suburbs to go food shopping. The market, in turn, is heavily stocked with local produce. "We estimated that a typical tomato traveled twenty-five hundred miles to reach our kitchen table," Clavelle said. "And we could produce that tomato right here." Finally, we headed over to a section of town known as the Intervale. A flood plain along the Winooski River, the Intervale once was a farming district, then it was a wasteland, and now it is home to an assortment of community gardens and cooperatives with names like the Lucky Ladies Egg Farm and the Stray Cat Farm. By the time we arrived at the Intervale, the weather had changed for the worse. In the pouring rain, we stopped at an old brick farmhouse. Summer squash of various shapes and sizes were displayed out front. Next door was a composting facility that collects vegetable waste from local restaurants and turns it back into soil.

"It's a closed loop," Clavelle told me.

One consequence, presumably unintended, of America's failure to ratify the Kyoto Protocol has been the emergence of a not-quite-grassroots movement. In February 2005, Greg Nickels, the mayor of Seattle, began to circulate a set of principles that he called the "U.S. Mayors Climate Protection Agreement." Within four months, more than a hundred and seventy mayors, representing some thirty-six million people, had signed on, including Mayor Michael Bloomberg of New York; Mayor John Hickenlooper of Denver; and Mayor Manuel Diaz of Miami. Signatories agreed to "strive to meet or beat the Kyoto Protocol targets in their own communities." At around the same time, officials from New York, New Jersey, Delaware, Connecticut, Massachusetts, Vermont, New Hampshire, Rhode Island, and Maine announced that they had reached a tentative agreement to freeze power plant emissions from their states at current levels and then begin to cut them. Even Governor Arnold Schwarzenegger, the Hummer collector, joined in; an executive order he signed in June 2005 called on California to reduce its greenhouse gas emissions to 2000 levels by 2010 and to 1990 levels by 2020. "I say the debate is over," Schwarzenegger declared right before signing the order.

Burlington's experience demonstrates how much can, indeed, be accomplished through local action. In the sixteen years since Clavelle became mayor, electricity usage in the state of Vermont has risen by nearly 15 percent. In Burlington, by contrast, it has dropped by one percent. The savings were achieved entirely through voluntary measures, by homeowners and businesses who, presumably, came to see controlling their utility bills as in their own self-interest.

But Burlington's experience also makes the limits of local action obvious. The biggest reductions were achieved early on, when the city approved a bond issue to fund energy conservation projects. As the most inefficient homes and businesses in the city were upgraded, gains became harder and harder to come by. Since the 10 percent challenge was initiated, in 2002, electricity demand in the city has actually started to creep back up again and is now slightly higher than it was at the campaign's launch. Meanwhile, whatever savings have been made in electricity usage have been offset by increased CO_2 emissions from other sources, mostly cars and trucks. As we were heading back to City Hall, I asked Clavelle what more could be done.

"It would be so much easier if we could say, 'Well, if we approved this one project or this action, the problem would be solved,'" he told me. "But there's no silver bullet. There's no one thing we can do. There's no *ten* things we can do. There's hundreds and hundreds of things that we need to do.

"I'm frustrated," he said. "But you need to remain hopeful."

The headquarters of the Natural Resources Defense Council are situated on West Twentieth Street in Manhattan. The offices, which occupy the top three floors of a twelve-story art deco building, were designed in 1989 as a prototype for energy-efficient urban life, with "occupancy sensors" that shut off the lights automatically when no one's around and special polymer-coated windows that help keep out heat. A large skylight above the staircase is supposed to provide natural light to the reception area, though after fifteen years the glass has been coated with a fine layer of New York grime.

David Hawkins runs NRDC's climate program. He is tall and thin, with dark, wavy hair and a gentle manner. Hawkins joined the environmental group thirty-five years ago, fresh out of law school, and has worked there ever since, with one break, in the late 1970s, when he served as head of the EPA's air quality division. These days, he spends a lot of his time in China, meeting with officials at places like the National Development and Reform Commission and the Shanxi Institute of Coal Chemistry.

Over the next fifteen years, the size of China's economy is expected to more than double. This projected growth, most of which will be fueled by coal, overwhelms not just all conservation projects that are currently being undertaken in the United States, but also any that could be reasonably imagined. Hawkins gave me a copy of a presentation he had prepared on future power plant construction. In it was a graph detailing China's plans: by 2010, the country is expected to build 150 new one thousand-megawatt coal plants (or their generating equivalent); by 2020, it is expected to construct another 168 new plants. If every single town and city in the United States were to match the efforts that Burlington has made, the aggregate savings would amount—very roughly—to 1.3 billion tons of carbon over the next several decades. Meanwhile, the lifetime emissions just from the new coal plants China is expected to build would amount to some 25 billion tons of carbon. To put this somewhat differently, China's new plants would burn through all of Burlington's savings—past, present, and future—in less than two and a half hours.

Despair might seem the logical response to such figures. In this way, the hazard of looking objectively at global warming can be almost as great as refusing to see the problem at all. Hawkins, though, is an optimist—perhaps by professional necessity. "If you're looking at global warming, you look at what the emissions are from the large industrial and industrializing countries," he told me. "And it doesn't take very long to conclude that you can't solve this problem unless you deal with the United States and China, and if you deal with the United States and China, you can solve this problem."

"China is in the takeoff stage," he went on. "So there's an opportunity to build things there using modern technology rather than to build them using pickup technology. And that's the challenge for us: to do things that convince the Chinese that that's the better strategy for them."

Right now, he pointed out, China is industrializing according to a model set in the United States forty or fifty years ago: its factories rely on obsolete and highly inefficient motors; its electricity transmission system is antiquated; and although it is the world's primary manufacturer of compact fluorescent bulbs, it barely uses any. (Per unit of gross domestic product, China consumes two and a half times as much energy as the United States and nearly nine times as much as Japan.) Were China to bring its factories up to date and fill even a modest amount of its projected energy demand from renewable sources, it is estimated that the number of new coal-fired plants it would need to build could be cut by nearly a third.

At this point, China is building only conventional coal-fired plants. For technical reasons, "carbon capture and storage," or CCS, isn't feasible with this type of plant. But if China were to shift to a method known as coal gasification, then—potentially at least—the CO_2 emissions from the new plants could be captured and sequestered. In that case, their carbon emissions would be substantially lower—possibly zero. It is estimated that together, coal gasification technology and carbon capture and storage would add 40 percent to the costs of a new plant. (This is an imprecise figure, since CCS has never actually been tried at a commercial power plant.) Hawkins has calculated that even assuming such a high differential cost, the added expense of carbon capture and storage for all the new coal plants expected to be built in all of the world's developing nations could be paid for through a one percent tax on the electricity bills of consumers in developed nations. "So it is affordable," he told me.

China's growth is often cited as a justification for U.S. inaction. What's the point of going to a lot of trouble if, in the end, your efforts won't make a difference? Hawkins maintains that this argument gets things completely backward. What America does, China in the long run will do too. "This isn't theory," he said. "We saw it with automobile pollution controls. We adopted those in the seventies and those modern pollution controls are being required around the world today. Sulfur dioxide scrubbers on power plants—we applied them; China is now applying them. There's a very practical reason why this works, and that is if a country like the United States embraces a cleanup technology, then the market starts to drive the price down, and other countries start to see that it is doable." Although no new coal-fired power plants have been built in the United States in recent years, many analysts expect this to change in the coming decade. Hawkins argues that American utilities should be prohibited from constructing any new plants without CCS capability.

"If we can get policies adopted that prevent the U.S. from building new coal plants that don't capture their emissions and create incentives for the Chinese to build new coal plants that will capture their emissions, then it doesn't matter if there's an international treaty or not," he said. "If we get the facts on the ground right, we've bought time."

Chapter 10

MAN IN THE ANTHROPOCENE

A FEW YEARS AGO, in an essay in *Nature*, the Nobel Prize–winning Dutch chemist Paul Crutzen coined a term. No longer, he wrote, should we think of ourselves as living in the Holocene. Instead, an epoch unlike any of those which preceded it had begun. This new age was defined by one creature—man—who had become so dominant that he was capable of altering the planet on a geological scale. Crutzen dubbed this age the "Anthropocene."

Crutzen's was not the first such neologism. Already in the 1870s, the Italian geologist Antonio Stoppani argued that human influence was ushering in a new age, which he called the "anthropozoic era." A few decades later, the Russian geochemist Vladimir Ivanovich Vernadsky proposed that the earth was entering a new stage—the "noosphere"—dominated by human thought. But while these earlier terms had had a positive slant—"I look forward with great optimism . . . We live in a transition to the noosphere," Vernadsky wrote—the connotations of the Anthropocene were distinctly cautionary. Humans had become the driving agents on the planet, yet it wasn't at all clear they knew where they were going.

Crutzen won the Nobel for his work on the chemistry of ozone depletion, a phenomenon that offers many parallels, both scientific and social, to global warming. The most prevalent ozone-destroying chemicals— chlorofluorocarbons—are odorless, colorless, nonreactive, and, much like CO_2, apparently benign. (To demonstrate their safety, their inventor once inhaled some CFCs and then used the vapors to blow out a set of birthday candles.) Starting in the 1930s, the "wonder gas" was employed as a refrigerant and in the 1940s as an ingredient in Styrofoam. The first indication that chlorofluorocarbons were any-thing to worry about didn't come until the 1970s, when research chemists began to consider—purely as an aca-demic exercise—what would happen to CFCs in the upper atmosphere. They determined that although the chemicals were stable near the earth's surface, in the stratosphere they wouldn't be. Once CFCs started to break down, the result would be free chlorine, which, they hypothesized, would work as a catalyst to convert ozone, O_3, into ordinary oxygen, O_2. Because stratospheric ozone shields the earth from ultraviolet radiation, the researchers warned that continued use of CFCs could have disastrous consequences. F. Sherwood Rowland, who shared the Nobel Prize with Crutzen, came home one night and told his wife, "The work is going well, but it looks like it might be the end of the world."

The damaging effects of CFCs were confirmed—rather

more dramatically than researchers had anticipated—in the 1980s by the discovery that a "hole" had opened up in the ozone layer over Antarctica. (Confirmation might have come earlier had NASA computers not been programmed to reject as erroneous any data on ozone levels that seemed too low.) Even as evidence of chlorofluorocarbons' effects accumulated, American chemical manufacturers, who supplied more than a third of the world's CFCs, continued to resist regulation, arguing on the one hand that more study of the problem was needed and on the other that only unified global action could address it. At one point, President Reagan's interior secretary, Donald Hodel, suggested that if CFCs were indeed destroying the ozone layer, then people should simply wear sunglasses and buy hats. "People who don't stand out in the sun—it doesn't affect them," he asserted. Finally, in 1987, the Montreal Protocol was agreed to, and the process of phasing out CFCs began. (Chlorofluorocarbons, it should be noted, are also a greenhouse gas.) It is expected that sometime in the next several years, ozone levels will bottom out and then begin to creep back up again. Depending on how you look at things, this resolution represents either a triumph of science, or just the reverse. As Crutzen himself has observed, if chlorine had turned out to behave just slightly differently in the upper atmosphere, or if its chemical cousin bromine had been used in its stead, then by the time anyone had thought to look into the state of the ozone layer, the "ozone hole" would have stretched from one pole to the other.

"More by luck than by wisdom this catastrophic situation did not develop," he has written.

In the case of global warming, a much longer interval separates theory and observation. According to Crutzen, the Anthropocene began all the way back in the 1780s, the decade in which James Watt perfected his steam engine. Arrhenius undertook his pen and paper calculations in the 1890s. The retreat of the Arctic sea ice, the warming of the oceans, the rapid shrinking of the glaciers, the redistribution of species, the thawing of the permafrost—these are all new phenomena. It is only in the last five or ten years that global warming has finally emerged from the background "noise" of climate variability. And even so, the changes that can be seen lag behind the changes that have been set in motion. The warming that has been observed so far is probably only about half the amount required to bring the planet back into energy balance. This means that even if carbon dioxide were to remain stable at today's levels, temperatures would still continue to rise, glaciers to melt, and weather patterns to change for decades to come.

But CO_2 levels are *not* going to remain stable. Just to slow the growth, as Socolow and Pacala's "wedge" scheme illustrates, is a hugely ambitious undertaking, one that would require new patterns of consumption, new technologies, and new politics. Whether the threshold for "dangerous anthropogenic interference" is 450 parts per million of CO_2 or 500, or even 550 or 600, the world is rapidly approaching the point at which, for all practical purposes, the crossing of that threshold will become impossible to prevent. To refuse to act, on the grounds that still more study is needed or that meaningful efforts are too costly or that they impose an unfair burden on industrialized nations, is not to put off the consequences, but to rush toward them. The British magazine *New Scientist* recently ran a global warming Q&A, which ended with the question, "How worried should we be?" The answer was another question: "How lucky do you feel?"

Luck and resourcefulness are, of course, essential human qualities. People are always imagining new ways to live, and then figuring out ways to remake the world to suit what they've imagined. This capacity has allowed us, collectively, to overcome any number of threats in the past, some imposed by nature and some by ourselves. It could be argued, taking this long view, that global warming will turn out to be just one more test in a sequence that already stretches from plague and pestilence to the prospect of nuclear annihilation. If, at this moment, the bind that we're in seems insoluble, once we've thought long and hard enough about it we'll find—or perhaps, float—our way clear.

But it's also possible to take an even longer view of the situation. The climate record provided by Greenland ice cores gives a highly resolved history going back more than a hundred thousand years and the Antarctic cores a history stretching back more than four hundred thousand years. What these records show, in addition to a clear correlation between CO_2 levels and global temperatures, is that the last glaciation was a time of frequent and traumatic climate swings. During that period, humans who were, genetically speaking, just like ourselves wandered the globe, producing nothing more permanent than isolated cave paintings and large piles of mastodon bones. Then, ten thousand years ago, the weather changed. As the climate settled down, so did we. People built villages, towns, and, finally, cities, along the way inventing all the basic technologies— agriculture, metallurgy, writing—that future civilizations would rely upon. These developments would not have been possible without human ingenuity, but, until the climate cooperated, ingenuity, it seems, wasn't enough.

Ice core records also show that we are steadily drawing closer to the temperature peaks of the last interglacial, when sea levels were some fifteen feet higher than they are today. Just a few degrees more and the earth will be hotter than it has been at any time since our species evolved. The feedbacks that have been identified in the climate system— the ice-albedo feedback, the water vapor feedback, the feedback between temperatures and carbon storage in the permafrost—take small changes to the system and amplify them into much larger forces. Perhaps the most unpredictable feedback of all is the human one. With six billion people on the planet, the risks are everywhere apparent. A disruption in monsoon patterns, a shift in ocean currents, a major drought—any one of these could easily produce streams of refugees numbering in the millions. As the effects of global warming become more and more difficult to ignore, will we react by finally fashioning a global response? Or will we retreat into ever narrower and more destructive forms of self-interest? It may seem impossible to imagine that a technologically advanced society could choose, in essence, to destroy itself, but that is what we are now in the process of doing.

Part III

TIME

Chapter 11

TEN YEARS ON

O N   M A Y   1 2 ,   2 0 1 4 , NASA scientists held a press conference to announce what they blandly described as "findings" on the West Antarctic Ice Sheet. Ice flows like water, only more slowly, and so the ice on Antarctica, like the ice on Greenland, is always on the move. The scientists had analyzed two decades' worth of satellite data to track the position of several enormous glaciers that flow into the Amundsen Sea. Then they'd used the data to map the terrain underneath these glaciers.

The scientists' findings were consistent. All the glaciers in the Amundsen Sea sector had retreated over the previous twenty years, one by nineteen miles, another by twenty-two miles, and a third by nine miles. The scientists had found no buried mountain ranges that might inhibit further retreat; on the contrary, the land under the glaciers, much of which lies below sea level, seemed to slope downward. Thus, they concluded, the retreat of the glaciers had begun to feed on itself. The more ice melted back, the more seawater would infiltrate under the ice sheet, and the farther back the ice would melt, until there was none left.

"Today we present observational evidence that a large sector of the West Antarctic Ice Sheet has gone into a state of irreversible retreat," Eric Rignot, of NASA's Jet Propulsion Laboratory in Pasadena, said, explaining the findings. "It has passed the point of no return."

The Amundsen Sea sector holds enough ice that, were it all to melt, global sea levels would rise by four feet. If this weren't bad enough, the scientists found that the loss of ice from the Amundsen Sea sector would likely destabilize adjacent sections of the ice sheet; this, Rignot warned, "could triple the contribution to sea level." The same day a second team, from the University of Washington, released a study of one of the most important West Antarctic glaciers, the Thwaites glacier; this study, too, concluded, that the ice sheet's disintegration was now inevitable. "West Antarctic Ice Sheet collapse is under way," stated the press release.

If accurate, the new research from Antarctica means that "dangerous anthropogenic interference" has now been achieved. Hundreds of millions—perhaps billions—of people currently live in cities that would be devastated by twelve feet of sea level rise. Though the process may take centuries to fully play out, what's key is that there's no going back.

"Scary," Stefan Rahmstorf, an oceanographer at Potsdam University, tweeted the day the findings were released. "One of the feared tipping points of the climate system appears to have been crossed."

The ice sheet studies made news for a couple of days.

The *New York Times*, which ran the story on the front page, observed that "the mighty West Antarctica ice sheet has begun falling apart." *Mother Jones* put it more pointedly. "This Is What a Holy Shit Moment for Global Warming Looks Like," it declared. Then it was back to business as usual. The same week the studies were published, Senator Marco Rubio, of Florida, stated on national TV that he did "not believe that human activity is causing these dramatic changes to our climate the way these scientists are portraying it." Rubio, a Republican, lives in Miami, which, it's worth noting, ranks as number one in the world among cities with assets vulnerable to sea level change. (In terms of population it ranks fourth on the list of vulnerable cities, just below Mumbai, Guangzhou, and Shanghai.) A few days later, after much criticism, Rubio shifted his stance. "I've never disputed that the climate is changing," he said. "I've pointed out that climate to some extent is always changing."

I did most of the reporting for this book in 2004. At that time the signs of global warming were clearly apparent to those who knew where to look for them. I wrote the book to enable those less immersed in the subject to see, if not exactly with their own eyes then at least through mine, the changes that were already under way. It seemed to me we were at a critical juncture and that people needed to recognize immediate action was necessary.

How have things changed since then? Well, for one thing, the signs of global warming have become much more obvious. They're less the sort you need to go looking for and more the sort you can't avoid. Across the American West, millions of acres of pine trees have died off, owing to warming-related beetle infestations. In the Southwest, forest fires are breaking out earlier, and burning more intensely. Superstorm Sandy hit the eastern United States in October 2012. While it's impossible to say whether the storm itself was a product of climate change, certainly Sandy, which caused an estimated sixty-five billion dollars in damage, was that much more destructive owing to higher sea levels.

Globally, new records for warmth were set in 2005 and then again in 2010. The ten hottest years on record have now all occurred since 1997. As I write this, California is in the midst of a historic dry spell, with all regions of the state classified as in either "serious," "extreme," or "exceptional" drought. Whenever I read a story about the drought—a recent piece in the *Los Angeles Times* chronicled the desperation of farm workers scrambling for work in the desiccated fields—I think of the projected drought maps that David Rind showed me when I was hanging out at GISS. And I recall what California's water managers told him when he presented these maps to them: "Well, if that happens, forget it."

The speed at which the world is changing has even caught many climate scientists by surprise. Consider the case of the Arctic sea ice. When I went to visit CRREL in 2004, the forecast was that the Arctic Ocean would be free of summer ice by the end of the century. The very next year, in 2005, the extent of the summer ice reached a new record low. In 2007, it hit another new low; in just two years its area had shrunk by an additional—and astonishing—23 percent. In the summer of 2012, the summer ice once again set a new record; the ice cap had contracted by an additional 18 percent.

"We are now in uncharted territory," Mark Serreze, director of the National Snow and Ice Data Center in Boulder, declared in 2012. While scientists had known for a long time that warming would be "seen first and be most pronounced in the Arctic," he said, "few of us were prepared for how rapidly the changes would actually occur." The perennial ice cap is now only about half

The extent of Arctic sea ice has declined much faster than climate models predicted. Credit: Julienne Stroeve, updated and adapted from Stroeve et al., Geophysical Research Letters, Vol. 34 (2007).

the size it was in the 1980s, and for the last few summers the fabled Northwest Passage has been mostly ice-free. In 2013, the Danish-operated Nordic Orion completed the passage, the first time in history that a large freighter had done so. And in 2014, *National Geographic* announced that it had had to make "drastic" changes to its *Atlas of the World* to reflect the shrinkage. It now seems likely that the perennial ice will be gone entirely by the middle of the century, if not sooner.

As the evidence of climate change has grown more apparent, so, too, has the need for action. And yet, by most measures, Americans remain unconvinced. A 2013 poll by Pew Research asked respondents to list their top priorities for Congress and the president. "Dealing with global warming" ranked nineteenth out of twenty possibilities, behind "reducing the influence of lobbyists" and "dealing with moral breakdown" and ahead of only "dealing with global trade." While two-thirds of Democrats now believe the climate is changing mostly due to human influence (once again according to Pew), fewer than half of Independents believe this and only a quarter of Republicans do.

Increasingly, the gap between the science of global warming and the public's response to it has itself become a subject of study. In the last few years, a whole new field has emerged devoted to explaining why the urgent warnings about climate change have gone ignored. Call it "clipsy." According to one school of thought, the explanation lies in the public's continuing confusion about what scientists are saying.

"The single most common myth about climate change among Americans is that there's a lot of disagreement among the experts," Ed Maibach, director of the Center for Climate Change Communication at George Mason University, told me. "And the reason why they think there is a lot of disagreement among the experts is because there was an intentional strategy to sow the seeds of doubt." Over the last few decades, dozens of scientific-*seeming* reports have purported to show there's no scientific consensus on climate change, even though the basic geophysics have been understood since Arrhenius. This deliberate misinformation campaign, as the historians of science Naomi Oreskes and Erik M. Conway have documented, was modeled on the tactics developed by the tobacco industry, and financed, sometimes directly and sometimes indirectly, by the fossil fuel industry.

According to a second school of thought, the problem lies not so much in how the science has been communicated but in how it's become tangled up with other issues.

"What you believe about climate change doesn't reflect what you know," said Dan Kahan, a professor at Yale Law School who studies risk perception. "It expresses who you are."

To illustrate this point, Kahan cited the results of yet another survey by the Pew Research Center. This survey was designed to test basic scientific knowledge and it posed questions like "What is the main function of red blood cells?" When respondents were asked what gas "most scientists believe causes temperatures in the atmosphere to rise," 58 percent chose the correct answer: "carbon dioxide." There was little difference in the proportion of Democrats and Republicans who gave the right response; among the former it was 56 percent, among the latter 58 percent. (Among Independents, 63 percent chose correctly.)

But polls that ask Americans about their own beliefs about global warming show a significant partisan divide; in another Pew survey, 66 percent of Democrats said they believed that human activity was the "main cause" of global warming, while only 24 percent of Republicans did. This suggests there are many Democrats who don't know what's causing climate change but still believe humans are responsible for it and many Republicans who *do* know, yet still deny that humans play a role. And what this shows, according to Kahan, is that people's views on climate change are shaped less by their knowledge of the science than by their sense of group identity. To break the political logjam, he argues, Americans need to find ways of talking about climate change that don't require members of one group or the other to renounce their cultural identity.

"If you show people there is some way of responding to the problem that's consistent with who they are, then they're more likely to see the problem," Kahan told me.

Kari Marie Norgaard is a sociologist at the University of Oregon who has studied how people talk about climate change. She, too, believes there's a strong cultural component to Americans' attitudes, but she sees the problem as reflecting the strategies people use to avoid painful subjects.

Norgaard argues that it's difficult even for people who are privately worried about climate change to discuss the issue in public because on the one hand they feel guilty about the situation and on the other they feel helpless to change it. "We have a need to think of ourselves as good people," she told me. Meanwhile, the very lack of discussion about the issue feeds itself: people feel that if it really were a serious problem, others would be dealing with it: "It's difficult for people to feel that climate change is really happening in part because we're embedded in a world where no one else around us is talking about it."

"It becomes a vicious cycle between the political gridlock and the cultural and individual gridlock," Norgaard went on.

What could possibly break this cycle? Norgaard argues that if the nation's political leaders would candidly discuss the issue "it could be very powerful. It could free up a lot of the hopelessness people feel and allow them to mobilize."

"I think there are probably multiple levels at which we could break this cycle," she went on. And though, after more than thirty years of ignored warnings, the challenge has grown all the more daunting, she said, "I don't believe we get to give up."

This book was originally divided into two parts, first "Nature," then "Man." The former covered the science of climate change, the latter our uncomfortable relationship to that science. The three chapters that follow were all written after the book came out and originally appeared as articles in the *New Yorker*. They do not fit neatly into either part, which is why they are being offered at the end of this edition. The new chapters are presented here as they originally appeared, and also in the order they were published. Thus, you will see certain figures, like the level of CO_2 in the atmosphere, increase as the years go by.

The first of these new chapters concerns ocean acidification, global warming's equally evil twin. Effectively, when we pump CO_2 into the air, we are also dissolving it in the sea, where it forms carbonic acid. As recently as a decade ago, even many marine biologists were unaware of ocean acidification; since then, it has risen very close to the top of the (long) list of threats to ocean life.

Owing to the extra CO_2 that's been added to the seas so far, the oceans are now about 30 percent more acidic than they were at the start of the Industrial Revolution. If current trends continue, by the end of this century they will be 150 percent more acidic. This is a huge change in what amounts to a geological instant, and it's anticipated that it will have drastic consequences, though what, exactly, those consequences will be is difficult to predict. (Ocean acidification is associated with some of the worst crises in the history of life, including the end–Permian extinction, 250 million years ago, which killed off something like 90 percent of all species then on the planet.)

Ocean acidification is apt to change the makeup of microbial communities, which, in turn, will alter the availability of crucial nutrients, like nitrogen and iron. It will change the way light passes through the water, and also sound. (It's expected the seas will become noisier.) Perhaps most significantly, it is going to make life more difficult for organisms like clams, oysters, and starfish that construct shells or external skeleton out of the mineral calcium carbonate. One group that appears particularly sensitive to changes in water chemistry is stony corals. Thus, a world of unchecked ocean acidification will also, most likely, be a world without coral reefs.

"The potential consequences of such acidification are nothing less than catastrophic," J.E.N. Veron, the former chief scientist of the Australian Institute of Marine Science, has written.

There was a time, not that long ago, when it seemed as if humanity's love affair with fossil fuels might come to an end not because we decided to break off the relationship, but because they did. So-called peak oilers warned that we had already burned through more than half of the world's readily accessible oil reserves and that we would soon be facing a series of dire shortages. But instead of running out, we have found ways to get at not-so-accessible sources of oil. The same holds true for natural gas.

In the second of the new chapters, I look at these "unconventional" fuels and their implications. The Canadian tar sands are one of the world's richest "unconventional" resources. As the name suggests, the tar sands are a mixture; they consist of clay and bits of quartzite and very heavy hydrocarbons known as bitumen. Tar sands can be strip-mined, like coal, a process that leaves behind an utterly ravaged landscape. Alternatively, the bitumen can be baked out of the ground using steam. It then can be diluted, sent through pipelines to refineries, and converted into what's called synthetic crude. The Canadian tar sands—a series of deposits that extend over (or, really, under) much of northern Alberta—hold enough bitumen to yield more than a trillion barrels of synthetic crude. Even if only 10 percent of this is actually recoverable, it still represents one of the largest oil deposits on the globe.

The turn to unconventional fuels—a turn made possible by new technologies like "steam assisted gravity drainage" and hydraulic fracturing, or "fracking"—has dramatically increased the reserves of fossil fuels that could, potentially, be combusted. Meanwhile, of course, the dynamics of the climate system have remained the same. This means that the gap between what we could burn and what we *should* has grown that much larger. In an influential article published in *Rolling Stone* in 2012, Bill McKibben laid out what he called "global warming's terrifying new math."

"We have five times as much oil and coal and gas on the books as climate scientists think is safe to burn," he wrote.

A criticism that's often leveled at environmentalists and, by extension, at environmental journalists is that they focus too much on the problems and not enough on the possible solutions. With this criticism in mind, a few years ago I visited the Danish island of Samsø, which sits in the Kattegat, an arm of the North Sea. The result is the third of the chapters that follow.

What drew me to Samsø was an unusual project spear-headed by a single, very energetic individual named Søren Hermansen. A sequence of rather idiosyncratic events had convinced Hermansen that Samsø could wean itself entirely off fossil fuels. Slowly, he managed to convince enough people on Samsø (population c. 4,000) the idea was worth pursuing that they began to convince the rest.

Samsingers, most of whom are farmers, invested in off-shore windmills and solar panels and tractors that could run on rapeseed oil. They competed with one another to see who could come up with more innovative ways to reduce their reliance on fossil fuels. Though the islanders did not completely give up fossil fuels—they still use gasoline in their cars—they did succeed in going "carbon neutral." They now produce more energy from renewable sources than, in total, they use. The Samsingers I spoke to were clearly proud of what they had done, but at the same time they were surprised to find their conservative farming community at the vanguard of social change. "We are only normal people," one dairy farmer told me. (The year after I profiled Hermansen, he was one of the co-winners of the Gothenburg Award for Sustainable Development, which means he shared a prize of a million Swedish crowns.)

For my part, I came away from Samsø both impressed by the islanders' commitment and saddened that I'd had to travel across the Atlantic to find a project like theirs. I'd like to think that the article gave people a sense of what is possible, and so perhaps helped to counter some of the feelings of powerlessness that surround climate change in this country. But I have yet to hear of an effort in the United States that's anywhere near as ambitious as Samsø's.

When Barack Obama took office in 2009, he vowed to act on climate change. In sharp contrast to his predecessor, George W. Bush, Obama appointed prominent scientists to key posts, including Jane Lubchenco, a marine ecologist, to head the National Oceanic and Atmospheric Administration; Steven Chu, a Nobel Prize-winning physicist to head the Energy Department; and John Holdren, a MacArthur-winning physicist, to be his chief science adviser. During his first term, the White House set ambitious new fuel efficiency standards for cars, and in his second term, the Environmental Protection Agency proposed regulations to cut carbon emissions from power plants by nearly 30 percent.

Yet Obama's record on climate change has, at best, been mixed, and overall, the political situation in the United States remains bleak. To achieve anything like the emissions reductions that climate scientists say are needed will require legislative action. During Obama's first year in office, it seemed such action was possible. In the spring of 2009, Democratic congressmen Ed Markey, of Massachusetts, and Henry Waxman, of California, pushed through the House a bill that would have instituted a "cap and trade" program for major emitters. (Under such a program, which was used, successfully, to reduce sulfur dioxide emissions, businesses like utilities and oil refineries would have to obtain permits to cover their CO_2 emissions; these permits could then be bought and sold, like commodities.) But the president put little political muscle behind the bill, and it died the following year in the Senate.

Since then, control of the House has switched to the Republicans, and many key positions are now occupied by lawmakers who—amazingly enough—still deny, or at least downplay, the risks of rising CO_2 levels. John Boehner, the House Speaker, has famously stated "the idea that carbon dioxide is a carcinogen that is harmful to our environment, is almost comical." Robert Aderholt, chairman of the Appropriations Committee's agriculture subcommittee, places himself in the "group of people who believe . . . that the earth is currently in a natural warming cycle," and Lamar Smith, head of the House Science Committee, recently stated that "the climate is changing due to a number of factors," including natural cycles. Over the last several years, no piece of climate legislation has been seriously considered by the House, and as I write this, the chances of such legislation being enacted in the foreseeable future seem to be effectively zero.

Meanwhile, at the international level the temporizing also continues. World leaders were supposed to hammer out a successor to the Kyoto Protocol in Copenhagen in 2009, but failed to do so. Instead, they came up with what was called the Copenhagen Accord, a nonbinding agreement that committed the world to trying to keep average global temperatures from rising by more than two degrees Celsius (3.6 Fahrenheit). In order to stay below two degrees Celsius, scientists generally agree, CO_2 levels cannot be allowed to climb above 450 parts per million (and even this may be too high). In 2009, global emissions fell slightly, owing to the financial crisis. But then they resumed their steady rise. They climbed to 8.3 billion metric tons in 2010, and to 9.4 billion metric tons by 2012. (Although the United States still ranks number one in terms of cumulative emissions, China is now the top emitter on an annual basis, having taken the lead in 2006, almost two decades earlier than anticipated.) Just to stabilize emissions at 2005 levels, then, the world would need not seven of Socolow and Pacala's wedges, as discussed earlier in this book, but nine or ten. A report from the World Bank warns that "present emission trends put the world plausibly on a path toward 4 degrees Celsius warming within the century." Many scientists and policy makers now believe that limiting warming to two degrees Celsius is probably for all intents and purposes impossible, and that it was already impossible at the time the limit was agreed to. The only way to "achieve the two-degree goal is to shut down the whole global economy," Yvo de Boer, the former executive director of the UN Framework Convention on Climate Change, said recently. In the spring of 2013, CO_2 levels reached 400 parts per million.

"We are in an unusual predicament," Vice President Al Gore once told Bill McKibben in an interview about climate change. "The maximum that is politically feasible, even the maximum that is politically imaginable right now, still falls short of the minimum that is scientifically and ecologically necessary."

In the years since I wrote this book I've been asked hundreds of variations on the question: "What should *I* do?" What people seem to be looking for is both advice on concrete actions they can take and the assurance that what they do will make a difference. Given the paralysis of the political system, the time lag built into the climate system, and the high likelihood that the threshold of DAI has now been crossed, it's difficult to offer such assurances. We have already changed the world dramatically, indeed quite probably catastrophically. But even when it comes to catastrophe, distinctions can be made. What we choose to do—or not to do—in the coming decades will determine the future both for our own kind and for the millions of other species with whom we share this planet. It is possible that we could still limit warming to *around* two degrees Celsius, and it is also possible that we could lock in warming of six degrees Celsius or more. These two possibilities represent radically different worlds.

Williamstown, Massachusetts

August 2014

Chapter 12

THE DARKENING SEA

PTEROPODS ARE TINY marine organisms that belong to the very broad class known as zooplankton.
Related to snails, they swim by means of a pair of winglike
gelatinous flaps and feed by entrapping even tinier marine
creatures in a bubble of mucus. Many pteropod species—
there are nearly a hundred in all—produce shells, apparently for protection; some of their predators, meanwhile,
have evolved specialized tentacles that they employ much
as diners use forks to spear escargot. Pteropods are first
male, but as they grow older they become female.

Victoria Fabry, an oceanographer at California State
University at San Marcos, is one of the world's leading
experts on pteropods. She is slight and soft-spoken, with
wavy black hair and blue-green eyes. Fabry fell in love
with the ocean as a teenager after visiting the Outer Banks,
off North Carolina, and took up pteropods when she was
in graduate school, in the early 1980s. At that point, most
basic questions about the animals had yet to be answered,
and, for her dissertation, Fabry decided to study their shell
growth. Her plan was to raise pteropods in tanks, but she

ran into trouble immediately. When disturbed, pteropods tend not to produce their mucus bubbles and slowly starve. Fabry tried using bigger tanks for her pteropods, but the only correlation, she recalled recently, was that the more time she spent improving the tanks, "the quicker they died." After a while, she resigned herself to constantly collecting new specimens. This, in turn, meant going out on just about any research ship that would have her.

Fabry developed a simple, if brutal, protocol that could be completed at sea. She would catch some pteropods, either by trawling with a net or by scuba diving, and place them in one-liter bottles filled with seawater, to which she had added a small amount of radioactive calcium-45. Forty-eight hours later, she would remove the pteropods from the bottles, dunk them in warm ethanol, and pull their bodies out with a pair of tweezers. Back on land, she would measure how much calcium-45 their shells had taken up during their two days of captivity.

In the summer of 1985, Fabry got a berth on a research vessel sailing from Honolulu to Kodiak Island. Late in the trip, near a spot in the Gulf of Alaska known as Station Papa, she came upon a profusion of *Clio pyramidata*, a half-inch-long pteropod with a shell the shape of an unfurled umbrella. In her enthusiasm, Fabry collected too many specimens; instead of putting two or three in a bottle, she had to cram in a dozen. The next day, she noticed that something had gone wrong. "Normally, their shells are transparent," she said. "They look like little gems, little jewels. They're just beautiful. But I could see that, along the edge, they were becoming opaque, chalky."

Like other animals, pteropods take in oxygen and give off carbon dioxide as a waste product. In the open sea, the CO_2 they produce has no effect. Seal them in a small container, however, and the CO_2 starts to build up, changing the water's chemistry. By overcrowding her *Clio pyramidata*, Fabry had demonstrated that the organisms were highly sensitive to such changes. Instead of growing, their shells were dissolving. It stood to reason that other kinds of pteropods—and, indeed, perhaps any number of shell-building species—were similarly vulnerable. This should have represented a major discovery, and a cause for alarm. But, as is so often the case with inadvertent breakthroughs, it went unremarked upon. No one on the boat, including Fabry, appreciated what the pteropods were telling them, because no one, at that point, could imagine the chemistry of an entire ocean changing.

Since the start of the industrial revolution, humans have burned enough coal, oil, and natural gas to produce some two hundred and fifty billion metric tons of carbon. The result, as is well known, has been a transformation of the earth's atmosphere. The concentration of CO_2 in the air today—380 parts per million—is higher than it has been at any point in the past eight hundred thousand years, and probably much longer. At the current rate of emissions growth, CO_2 concentration will top 500 parts per million—roughly double preindustrial levels—by the middle of this century. It is expected that such an increase will prompt a string of disasters, including fiercer hurricanes, more deadly droughts, the disappearance of most remaining glaciers, the melting of the Arctic ice cap, and the inundation of many of the world's major coastal cities. But this is only half the story.

Ocean covers 70 percent of the earth's surface, and everywhere that water and air come into contact there is an exchange. Gases from the atmosphere get absorbed by the ocean and gases dissolved in the water are released into the atmosphere. When the two are in equilibrium, roughly the same quantities are being dissolved as are getting released. But change the composition of the atmosphere, as we have done, and the exchange becomes lopsided: more CO_2 from the air enters the water than comes back out. In the 1990s, researchers from seven countries conducted nearly a hundred cruises, and collected more than seventy thousand seawater samples from different depths and locations. The analysis of these samples, which was completed in 2004, showed that nearly half of all the carbon dioxide that humans have emitted since the start of the nineteenth century has been absorbed by the sea.

When CO_2 dissolves, it produces carbonic acid, which has the chemical formula H_2CO_3. As acids go, H_2CO_3 is relatively innocuous—we drink it all the time in Coke and other carbonated beverages—but in sufficient quantities it can change the water's pH. Already, humans have pumped enough carbon into the oceans—some hundred and twenty billion tons—to produce a .1 decline in surface pH. Since pH, like the Richter scale, is a logarithmic measure, a .1 drop represents a rise in acidity of about 30 percent. The process is generally referred to as "ocean acidification," though it might more accurately be described as a decline in ocean alkalinity. This year alone, the seas will absorb an additional two billion tons of carbon, and next year it is expected that they will absorb another two billion tons. Every day, every American, in effect, adds seven pounds of carbon to the oceans.

Because of the slow pace of deep-ocean circulation and the long life of carbon dioxide in the atmosphere, it is impossible to reverse the acidification that has already taken place. Nor is it possible to prevent still more from occurring. Even if there were some way to halt the emission of CO_2 tomorrow, the oceans would continue to take up carbon until they reached a new equilibrium with the air. As Britain's Royal Society noted in a 2005 report, it will take "tens of thousands of years for ocean chemistry to return to a condition similar to that occurring at pre-industrial times."

Humans have, in this way, set in motion change on a geologic scale. The question that remains is how marine life will respond. Though oceanographers are just beginning to address the question, their discoveries, at this early stage, are disturbing. A few years ago, Fabry finally pulled her cloudy shells out of storage to examine them with a scanning electron microscope. She found that their surfaces were riddled with pits. In some cases, the pits had grown into gashes, and the upper layer had started to pull away, exposing the layer underneath.

The term "ocean acidification" was coined in 2003 by two climate scientists, Ken Caldeira and Michael Wickett, who were working at the Lawrence Livermore National Laboratory, in Northern California. Caldeira has since moved to the Carnegie Institution, on the campus of Stanford University, and during the summer of 2006 I went to visit him at his office, which is housed in a "green" building that looks like a barn that's been taken apart and reassembled at odd angles. The building has no air-conditioning; temperature control is provided by a shower of mist that rains down into a tiled chamber in the lobby. At the time of my visit, California was in the midst of a record-breaking heat wave; the system worked well enough that Caldeira's office, if not exactly cool, was at least moderately comfortable.

Caldeira is a trim man with wiry brown hair and a boyish sort of smile. In the 1980s, he worked as a software developer on Wall Street, and one of his clients was the New York Stock Exchange, for whom he designed computer programs to help detect insider trading. The programs functioned as they were supposed to, but after a while Caldeira came to the conclusion that the NYSE wasn't actually interested in catching insider traders, and he decided to switch professions. He went back to school, at New York University, and ended up becoming a climate modeler.

Unlike most modelers, who focus on one particular aspect of the climate system, Caldeira is, at any given moment, working on four or five disparate projects. He particularly likes computations of a provocative or surprising nature; for example, not long ago he calculated that cutting down all the world's forests and replacing them with grasslands would have a slight cooling effect. (Grasslands, which are lighter in color than forests, absorb less sunlight.) Other recent calculations that Caldeira has made show that to keep pace with the present rate of temperature change plants and animals would have to migrate poleward by thirty feet a day, and that a molecule of CO_2 generated by burning fossil fuels will, in the course of its lifetime in the atmosphere, trap a hundred thousand times more heat than was released in producing it.

Caldeira began to model the effects of carbon dioxide on the oceans in 1999, when he did some work for the Department of Energy. The department wanted to know what the environmental consequences would be of capturing CO_2 from smokestacks and injecting it deep into the sea. Caldeira set about calculating how the ocean's pH would change as a result of deep-sea injection and then compared that result with the current practice of pouring carbon dioxide into the atmosphere and allowing it to be taken up by surface waters. In 2003, he submitted his work to *Nature*. The journal's editors advised him to drop the discussion of deep-ocean injection, he recalled, because the calculations concerning the effects of ordinary atmospheric release were so startling. Caldeira published the first part of his paper under the subheading "The coming centuries may see more ocean acidification than the past 300 million years."

Caldeira told me that he had chosen the term "ocean acidification" quite deliberately, for its shock value. Seawater is naturally alkaline, with a pH ranging from 7.8 to 8.5—a pH of 7 is neutral—which means that, for now, at least, the oceans are still a long way from actually turning acidic. Meanwhile, from the perspective of marine life, the drop in pH matters less than the string of chemical reactions that follow.

The main building block of shells is calcium carbonate—$CaCO_3$. (The White Cliffs of Dover are a huge $CaCO_3$ deposit, the remains of countless tiny sea creatures that piled up during the Cretaceous—or "chalky"—period.) Calcium carbonate produced by marine organisms comes in two principal forms, aragonite and calcite, which have slightly different crystal structures. How, exactly, different organisms form calcium carbonate remains something of a mystery. Ordinarily in seawater, $CaCO_3$ does not precipitate out as a solid. To build their shells, calcifying organisms must, in effect, assemble it. Adding carbonic acid to the water complicates their efforts because it reduces the number of carbonate ions in circulation. In scientific terms, this is referred to as "lowering the water's saturation state with respect to calcium carbonate." Practically, it means shrinking the supply of material available for shell formation. (Imagine trying to build a house when someone keeps stealing your bricks.) Once the carbonate concentration gets

FIELD NOTES FROM A CATASTROPHE

pushed low enough, even existing shells, like those of Fabry's pteropods, begin to dissolve.

To illustrate, in mathematical terms, what the seas of the future will look like, Caldeira pulled out a set of graphs. Plotted on one axis was aragonite saturation levels; on the other, latitude. (Ocean latitude is significant because saturation levels tend naturally to decline toward the poles.) Different colors of lines represented different emissions scenarios. Some scenarios project that the world's economy will continue to grow rapidly and that this growth will be fueled mostly by oil and coal. Others assume that the economy will grow more slowly, and still others that the energy mix will shift away from fossil fuels. Caldeira considered four much-studied scenarios, ranging from one of the most optimistic, known by the shorthand B1, to one of the most pessimistic, A2. The original point of the graphs was to show that each scenario would produce a different ocean. But they turned out to be more similar than Caldeira had expected.

Under all four scenarios, by the end of this century the waters around Antarctica will become undersaturated with respect to aragonite—the form of calcium carbonate produced by pteropods and corals. (When water becomes undersaturated, it is corrosive to shells.) Meanwhile, surface pH will drop by another .2, bringing acidity to roughly double what it was in preindustrial times. To look still further out into the future, Caldeira modeled what would happen if humans burned through all the world's remaining fossil fuel resources, a process that would

release some eighteen thousand gigatons of carbon dioxide. He found that by 2300 the oceans would become
undersaturated from the poles to the equator. Then he
modeled what would happen if we pushed still further and
burned through unconventional fuels, like low-grade
shales. In that case, we would drive the pH down so
low that the seas would come very close to being acidic.

"I used to think of B1 as a good scenario, and I used to
think of A2 as a terrible scenario," Caldeira told me. "Now
I look at them as different flavors of bad scenarios."

He went on. "I think there's a whole category of
organisms that have been around for hundreds of millions
of years which are at risk of extinction—namely, things
that build calcium-carbonate shells or skeletons. To a first
approximation, if we cut our emissions in half it will take
us twice as long to create the damage. But we'll get to
more or less the same place. We really need an order-of-
magnitude reduction in order to avoid it."

Caldeira said that he had recently gone to Washington
to brief some members of Congress. "I was asked, 'What is
the appropriate stabilization target for atmospheric CO_2?'"
he recalled. "And I said, 'Well, I think it's inappropriate to
think in terms of stabilization targets. I think we should
think in terms of emissions targets.' And they said, 'Okay,
what's the appropriate emissions target?' And I said, 'Zero.'

"If you're talking about mugging little old ladies, you
don't say, 'What's our target for the rate of mugging little
old ladies?' You say, 'Mugging little old ladies is bad, and
we're going to try to eliminate it.' You recognize you

might not be a hundred percent successful, but your goal is to eliminate the mugging of little old ladies. And I think we need to eventually come around to looking at carbon-dioxide emissions the same way."

Coral reefs grow in a great swath that stretches like a belt around the belly of the earth, from thirty degrees north to thirty degrees south latitude. The world's largest reef is the Great Barrier Reef, off the coast of northeastern Australia, and the second largest is off the coast of Belize. There are extensive coral reefs in the tropical Pacific, in the Indian Ocean, and in the Red Sea, and many smaller ones in the Caribbean. These reefs, home to an estimated 25 percent of all marine fish species, represent some of the most diverse ecosystems on the planet.

Much of what is known about coral reefs and ocean acidification was originally discovered, improbably en-ough, in Arizona, in the self-enclosed, supposedly self-sufficient world known as Biosphere 2. A three-acre glassed-in structure shaped like a ziggurat, Biosphere 2 was built in the late 1980s by a private group—a majority of the funding came from the billionaire Edward Bass—and was intended to demonstrate how life on earth (Biosphere 1) could be re-created on, say, Mars. The building contained an artificial "ocean," a "rain forest," a "desert," and an "agricultural zone." The first group of Biosphereans—four men and four women—managed to remain, sealed inside, for two years. They produced all their own food and, for a long stretch, breathed only recycled air, but the project was widely considered a failure. The Biosphereans spent much of the time hungry, and, even more ominously, they lost control of their artificial atmosphere. In the various "ecosystems," decomposition, which takes up oxygen and gives off CO_2, was supposed to be balanced by photosynthesis, which does the reverse. But, for reasons mainly having to do with the richness of the soil that had been used in the "agricultural zone," decomposition won out. Oxygen levels inside the building kept falling, and the Biosphereans developed what amounted to altitude sickness. Carbon dioxide levels soared, at one point reaching 3,000 parts per million, or roughly eight times the levels outside.

When Biosphere 2 officially collapsed, in 1995, Columbia University took over the management of the building. The university's plan was to transform it into a teaching and research facility, and it fell to a scientist named Chris Langdon to figure out something pedagogically useful to do with the "ocean," a tank the size of an Olympic swimming pool. Langdon's specialty was measuring photosynthesis, and he had recently finished a project, financed by the U.S. Navy, that involved trying to figure out whether blooms of bioluminescent algae could be used to track enemy submarines. (The answer was no.) Langdon was looking for a new project, but he wasn't sure what the "ocean" was good for. He began by testing various properties of the water. As would be expected in such a high-CO_2 environment, he found that the pH was low.

"The very first thing I did was try to establish normal chemistry," he recalled recently. "So I added chemicals— essentially baking soda and baking powder—to the water to bring the pH back up." Within a week, the alkalinity had dropped again, and he had to add more chemicals. The same thing happened. "Every single time I did it, it went back down, and the rate at which it went down was proportional to the concentration. So, if I added more, it went down faster. So I started thinking, What's going on here? And then it dawned on me."

Langdon left Columbia in 2004 and now works at the Rosenstiel School of Marine and Atmospheric Science, at the University of Miami. He has a high forehead, deep-set blue eyes, and a square chin. When I went to visit him, he took me to see his coral samples, which were growing in a sort of aquatic nursery across the street from his office. On the way, we had to pass through a room filled with tanks of purple sea slugs, which were being raised for medical research. In the front row, the youngest sea slugs, about half an inch long, were floating gracefully, as if suspended in gelatin. Toward the back were slugs that had been fed for several months on a lavish experimental diet. These were the size of my forearm and seemed barely able to lift their knobby, purplish heads.

Langdon's corals were attached to tiles arranged at the bottom of long, sinklike tanks. There were hundreds of them, grouped by species: *Acropora cervicornis*, a type of staghorn coral that grows in a classic antler shape; *Mon-*

tastrea cavernosa, a coral that looks like a seafaring cactus; and *Porites divaricata*, a branching coral made up of lumpy, putty-colored protuberances. Water was streaming into the tanks, but when Langdon put his hand in front of the faucet to stop the flow, I could see that every lobe of *Porites divaricata* was covered with tiny pink arms and that every arm ended in soft, fingerlike tentacles. The arms were waving in what looked to be a frenzy either of joy or of supplication.

Langdon explained that the arms belonged to separate coral polyps, and that a reef consisted of thousands upon thousands of polyps spread, like a coating of plaster, over a dead calcareous skeleton. Each coral polyp is a distinct individual, with its own tentacles and its own digestive system, and houses its own collection of symbiotic algae, known as zooxanthellae, which provide it with most of its nutrition. At the same time, each polyp is joined to its neighbors through a thin layer of connecting tissue, and all are attached to the colony's collective skeleton. Individual polyps constantly add to the group skeleton by combining calcium and carbonate ions in a medium known as the extracytoplasmic calcifying fluid. Meanwhile, other organisms, like parrot fish and sponges, are constantly eating away at the reef in search of food or protection. If a reef were ever to stop calcifying, it would start to shrink and eventually would disappear.

"It's just like a tree with bugs," Langdon explained. "It needs to grow pretty quickly just to stay even."

As Langdon struggled, unsuccessfully, to control the

pH in the Biosphere "ocean," he started to wonder whether the corals in the tank might be to blame. The Biosphereans had raised twenty different species of coral, and while many of the other creatures, including nearly all the vertebrates selected for the project, had died out, the corals had survived. Langdon wondered whether the chemicals he was adding to raise the pH were, by increasing the saturation state, stimulating their growth. At the time, it seemed an unlikely hypothesis, because the prevailing view among marine biologists was that corals weren't sensitive to changes in saturation. (In many textbooks, the formula for coral calcification is still given incorrectly, which helps explain the prevalence of this view.) Just about everyone, including Langdon's own postdoc, a young woman named Francesca Marubini, thought that his theory was wrong. "It was a total pain in the ass," Langdon recalled.

To test his hypothesis, Langdon employed a straightforward but time-consuming procedure. Conditions in the "ocean" would be systematically varied, and the growth of the coral monitored. The experiment took more than three years to complete, produced more than a thousand measurements, and, in the end, confirmed Langdon's hypothesis. It revealed a more or less linear relationship between how fast the coral grew and how highly saturated the water was. By proving that increased saturation spurs coral growth, Langdon also, of course, demonstrated the reverse: when saturation drops, coral growth slows. In the artificial world of Biosphere 2, the implications of this discovery were interesting; in the real world they were rather grim. Any drop in the ocean's saturation levels, it seemed, would make coral more vulnerable.

Langdon and Marubini published their findings in the journal *Global Biogeochemical Cycles* in the summer of 2000. Still, many marine biologists remained skeptical, in no small part, it seems, because of the study's association with the discredited Biosphere project. In 2001, Langdon sold his house in New York and moved to Arizona. He spent another two years redoing the experiments, with even stricter controls. The results were essentially identical. In the meantime, other researchers launched similar experiments on different coral species. Their findings were also the same, which, as Langdon put it to me, "is the best way to make believers out of people."

Coral reefs are under threat for a host of reasons: bottom trawling, dynamite fishing, coastal erosion, agricultural runoff, and, nowadays, global warming. When water temperatures rise too high, corals expel the algae that nourish them. (The process is called "bleaching," because without their zooxanthellae corals appear white.) For a particular reef, any one of these threats could potentially be fatal. Ocean acidification poses a different kind of threat, one that could preclude the very possibility of a reef.

Saturation levels are determined using a complicated formula that involves multiplying the calcium and carbonate ion concentrations, and then dividing the result by a figure called the stoichiometric solubility product. Prior to the industrial revolution, the world's major reefs were all growing in water whose aragonite saturation level stood between 4 and 5. Today, there is not a single remaining region in the oceans where the saturation level is above 4.5, and there are only a handful of spots—off the northeastern coast of Australia, in the Philippine Sea, and near the Maldives—where it is above 4. Since the take-up of CO_2 by the oceans is a highly predictable physical process, it is possible to map the saturation levels of the future with great precision. Assuming that current emissions trends continue, by 2060 there will be no regions left with a level above 3.5. By 2100, none will remain above 3.

As saturation levels decline, the rate at which reefs add aragonite through calcification and the rate at which they lose it through bioerosion will start to approach each other. At a certain point, the two will cross, and reefs will begin to disappear. Precisely where that point lies is difficult to say, because erosion may well accelerate as ocean pH declines. Langdon estimates that the crossing point will be reached when atmospheric CO_2 levels exceed 650 parts per million, which, under a "business as usual" emissions scenario, will occur sometime around 2075.

"I think that this is just an absolute limit, something they can't cope with," he told me. Other researchers put the limit somewhat higher, and others somewhat lower.

Meanwhile, as global temperatures climb, bleaching events are likely to become more common. A major worldwide bleaching event occurred in 1998, and in 2002, there was another event on the Great Barrier Reef.

Caribbean reefs suffered from a major bleaching event again in the summer of 2005. Taken together, acidification and rising ocean temperatures represent a kind of double bind for reefs: regions that remain hospitable in terms of temperature are becoming increasingly inhospitable in terms of saturation, and vice versa.

"While one, bleaching, is an acute stress that's killing them off, the other, acidification, is a chronic stress that's preventing them from recovering," Joanie Kleypas, a reef scientist at the National Center for Atmospheric Research, in Boulder, Colorado, told me. Kleypas said she thought that some corals would be able to migrate to higher latitudes as the oceans warm, but that, because of the lower saturation levels, as well as the difference in light regimes, the size of these migrants would be severely limited. "There's a point where you're going to have coral but no reefs," she said.

The tropical oceans are, as a rule, nutrient-poor; they are sometimes called liquid deserts. Reefs are so dense with life that they are often compared to rain forests. This rain-forest-in-the-desert effect is believed to be a function of a highly efficient recycling system, through which nutrients are, in effect, passed from one reef-dwelling organism to another. It is estimated that at least a million, and perhaps as many as nine million, distinct species live on or near reefs.

"Being conservative, let's say it's a million species that live in and around coral," Ove Hoegh-Guldberg, an expert on coral reefs at the University of Queensland, in Australia, told me. "Some of these species that hang around coral reefs can sometimes be found living without coral. But most species are completely dependent on coral—they literally live in, eat, and breed around coral. And, when we see coral get destroyed during bleaching events, those species disappear. The key question is how vulnerable all these various species are. That's a very important question, but at the moment you'd have to say that a million different species are under threat."

He went on. "This is a matter of the utmost importance. I can't really stress it in words strong enough. It's a do-or-die situation."

Around the same time that Langdon was performing his coral experiments at the Biosphere, a German marine biologist named Ulf Riebesell decided to look into the behavior of a class of phytoplankton known as coccolithophores. Coccolithophores build plates of calcite—coccoliths—that they arrange around themselves, like armor, in structures known as coccospheres. (Viewed under an electron microscope, they look like balls that have been covered with buttons.) Coccolithophores are very tiny—only a few microns in diameter—and also very common. One of the species that Riebesell studied, *Emiliani huxleyi*, produces blooms that can cover forty thousand square miles, turning vast sections of the ocean an eerie, milky blue.

In his experiments, Riebesell bubbled CO_2 into tanks of coccolithophores to mimic the effects of rising atmospheric concentrations. Both of the species he was studying—

Emiliani huxleyi and *Gephyrocapsa oceanica*—showed a clear response to the variations. As CO_2 levels rose, not only did the organisms' rate of calcification slow; they also started to produce deformed coccoliths and ill-shaped coccospheres.

"To me, it says that we will have massive changes," Riebesell, who works at the Leibniz Institute of Marine Sciences, in Kiel, told me. "If a whole group of calcifiers drops out, are there other organisms taking their place? What is the rate of evolution to fill those spaces? That's awfully difficult to address in experimental work. These organisms have never, ever seen this in their entire evolutionary history. And if they've never seen it they probably will find it difficult to deal with."

Calcifying organisms come in a fantastic array of shapes, sizes, and taxonomic groups. Echinoderms like starfish are calcifiers. So are mollusks like clams and oysters, and crustaceans like barnacles, and many species of bryozoans, or sea mats, and tiny protists known as foraminifera—the list goes on and on. Without experimental data, it's impossible to know which species will prove to be particularly vulnerable to declining pH and which will not. In the natural world, the pH of the water changes by season, and even time of day, and many species may be able to adapt to new conditions, at least within certain bounds. Obviously, though, it's impractical to run experiments on tens of thousands of different species. (Only a few dozen have been tested so far.) Meanwhile, as the example of coral reefs makes clear, what's more important than how acidification will affect any particular organism is how it will affect entire marine ecosystems—a question that can't be answered by even the most ambitious experimental protocol. The 2005 report by Britain's Royal Society noted that it was "not possible to predict" how whole communities would respond, but went on to observe that "without significant action to reduce CO_2 emissions" there may be "no place in the future oceans for many of the species and ecosystems we know today."

Carol Turley is a senior scientist at Plymouth Marine Laboratory, in Plymouth, England, and one of the authors of the Royal Society report. She observed that pH is a critical variable not just in calcification but in other vital marine processes, like the cycling of nutrients.

"It looks like we'll be changing lots of levels in the food chain," Turley told me. "So we may be affecting the primary producers. We may be affecting larvae of zoo-plankton and so on. What I think might happen, and it's pure speculation, is that you may get a shortening of the food chain so that only one or two species comes out on top—for instance, we may see massive blooms of jellyfish and things like that, and that's a very short food chain."

Thomas Lovejoy, who coined the term "biodiversity" in 1980, compared the effects of ocean acidification to "running the course of evolution in reverse."

"For an organism that lives on land, the two most important factors are temperature and moisture," Lovejoy, who teaches at George Mason University, told me. "And for an organism that lives in the water the two most important factors are temperature and acidity. So this is just a profound, profound change. It is going to send all kinds of ripples through marine ecosystems, because of the importance of calcium carbonate for so many organisms in the oceans, including those at the base of the food chain. If you back off and look at it, it's as if you or I went to our annual physical and the body chemistry came back and the doctor looked really, really worried. It's a systemic change. You could have food chains collapse, and fisheries ultimately with them, because most of the fish we get from the ocean are at the end of long food chains. You probably will see shifts in favor of invertebrates, or the reign of jellyfish."

Riebesell put it this way: "The risk is that at the end we will have the rise of slime."

Paleooceanographers study the oceans of the geologic past. For the most part, they rely on sediments pulled up from the bottom of the sea, which contain what might be thought of as a vast library written in code. By analyzing the oxygen isotopes of ancient shells, paleooceanographers can, for example, infer the temperature of the oceans going back at least a hundred million years, and also determine how much—or how little—of the planet was covered by ice. By analyzing mineral grains and deposits of "microfossils," they can map archaic currents and wind patterns, and by examining the remains of foraminifera they can re-create the history of ocean pH.

In September 2006, two dozen paleooceanographers met with a roughly equal number of marine biologists at a conference hosted by Columbia University's Lamont-

Doherty Earth Observatory. The point of the conference, which was titled "Ocean Acidification—Modern Observations and Past Experiences," was to use the methods of paleooceanography to look into the future. (The ocean-acidification community is still a relatively small one, and at the conference I ran into half the people I had spoken to about the subject, including Victoria Fabry, Ken Caldeira, and Chris Langdon.) Most of the meeting's first day was devoted to a discussion of an ecological crisis known as the Paleocene-Eocene Thermal Maximum, or PETM.

The PETM took place fifty-five million years ago, at the border marking the end of the Paleocene epoch and the beginning of the Eocene, when there was a sudden, enormous release of carbon into the atmosphere. After the release, temperatures around the world soared; the Arctic, for instance, warmed by ten degrees Fahrenheit, and Antarctica became temperate. Presumably because of this, vertebrate evolution veered off in a new direction. Many of the so-called archaic mammals became extinct and were replaced by entirely new orders: the ancestors of today's deer, horses, and primates all appeared right around the time of the PETM. The members of these new orders were curiously undersized—the earliest horse was no bigger than a poodle—a function, it is believed, of hot, dry conditions that favored smallness.

In the oceans, temperatures rose dramatically and, because of all the carbon, the water became increasingly acidic. Marine sediments show that many calcifying organisms vanished—more than fifty species of foraminifera, for example, died out—while others that were once rare became dominant. On the seafloor, the usual buildup of empty shells from dead calcifiers ceased. In ocean cores, the PETM shows up vividly as a band of reddish clay sandwiched between thick layers of calcium carbonate.

No one is sure exactly where the carbon of the PETM. came from or what triggered its release. (Deposits of natural gas known as methane hydrates, which sit, frozen, underneath the ocean floor, are one possible source.) In all, the release amounted to about two trillion metric tons, or eight times as much carbon as humans have added to the atmosphere since industrialization began. This is obviously a significant difference in scale, but the consensus at the conference was that if there was any disparity between then and now it was that the impact of the PETM was not drastic enough.

The seas have a built-in buffering capacity: if the water's pH starts to drop, shells and shell fragments that have been deposited on the ocean floor begin to dissolve, pushing the pH back up again. This buffering mechanism is highly effective, provided that acidification takes place on the same timescale as deep-ocean circulation. (One complete exchange of surface and bottom water takes thousands of years.) Paleooceanographers estimate that the release of carbon during the PETM took between one and ten thousand years—the record is not detailed enough to be more exact—and thus occurred too rapidly to be completely buffered. Currently, CO_2 is being released into the air at least three times and perhaps as much as thirty times as quickly as during the PETM. This is so fast that buffering by ocean sediments is not even a factor.

"In our case, the surface layer is bearing all the burden," James Zachos, a paleooceanographer at the University of California at Santa Cruz, told me. "If anything, you can look at the PETM as a best-case scenario." Ken Caldeira said that he thought a better analogy for the future would be the so-called K-T, or Cretaceous-Tertiary, boundary event, which occurred sixty-five million years ago, when an asteroid six miles wide hit the earth. In addition to dust storms, fires, and tidal waves, the impact is believed to have generated huge quantities of sulfuric acid.

"The K-T boundary event was more extreme but shorter-lived than what we could do in the coming centuries," Caldeira said. "But by the time we've burned conventional fossil fuel resources what we've done will be comparable in extremeness, except that it will last millennia instead of years." More than a third of all marine genera disappeared at the K-T boundary. Half of all coral species became extinct, and it took the other half more than two million years to recover.

Ultimately, the seas will absorb most of the CO_2 that humans emit. (Over the very long term, the figure will approach 90 percent.) From a certain vantage point, this is a lucky break. Were the oceans not providing a vast carbon sink, almost all of the CO_2 that humans have emitted would still be in the air. Atmospheric concentrations would now be nearing 500 parts per million, and the disasters predicted for the end of the century would already be upon us. That there is still a chance to do something to avert the worst consequences of global warming is thanks largely to the oceans.

But this sort of accounting may be misleading. As the process of ocean acidification demonstrates, life on land and life in the seas can affect each other in unexpected ways. Actions that might appear utterly unrelated—say, driving a car down the New Jersey Turnpike and secreting a shell in the South Pacific—turn out to be connected. To alter the chemistry of the seas is to take a very large risk, and not just with the oceans.

Originally published November 2006

Chapter 13

UNCONVENTIONAL CRUDE

THE TOWN OF FORT McMurray occupies a set of irregularly spaced hillsides on either side of the Athabasca River, in northern Alberta. It has a dozen check-cashing joints, a roughly equal number of hotels, and a gaming center called the Boomtown Casino. It also has a museum, which is devoted to the region's most important resource, the Alberta tar sands. Exhibits include an eight-foot-long rotor, half of a hundred-and-fifty-ton truck, and a pump of Brobdingnagian proportions. Near the entrance to the museum sits a black mound covered by a clear plastic dome. A sign invites visitors to scratch around in the mound with a little retractable rake, then lift up a flap and take a sniff. Tar sands look like dirt and smell like diesel fuel.

The tar sands begin near the border of Saskatchewan, around the latitude of Edmonton, and extend, in three major deposits, north and west almost to British Columbia. All in all, they cover—or, more accurately, underlie—some fifty-seven thousand square miles, an area roughly the size of Florida. It is believed that they were pushed into their present location seventy million years ago by the uplift of the Rocky Mountains.

For the most part, the tar sands consist of quartzite, clay, and water. The other ingredient—the "tar"—is a mixture of very heavy hydrocarbons known as bitumen. Bitumen can be used as a sealant—supposedly the word "mummy" is derived from the term in ancient Persian—and as a paving material. With the right technology, it can also be converted into a form of petroleum known as synthetic crude.

There are two ways to assess the world's oil supply. One is to consider only conventional reserves—the sort of oil that comes gushing out of the ground. Estimates of conventional reserves vary widely, but most analyses suggest that their output will begin to decline sometime in the next few decades (if it hasn't already)—a development that so-called peak oilers predict will lead to a variety of gruesome consequences, including blackouts, food shortages, and general economic collapse. The second way is to look beyond conventional reserves to unconventional ones, like the tar sands.

It is estimated that there is enough bitumen in Alberta to yield 1.7 trillion barrels of synthetic crude. Assuming that only 10 percent of this is actually recoverable, it still represents the second-largest oil reserve in the world, after Saudi Arabia's, and more oil than is contained in the reserves of Kuwait, Norway, and Russia put together. Unconventional crude can be found in many other parts of the globe besides Canada; these include eastern Venezuela, which is home to a huge tar-sands-like deposit called the Faja Petrolífera del Orinoco, and portions of Colorado, Utah, and Wyoming, where there's a thick layer of oil shale known as the Green River Formation. Even coal can be converted into liquid fuel. During the Second World War, the Nazis employed a technique called the Fischer-Tropsch process; the same process is now in use in several countries, most notably South Africa, which invested heavily in coal-to-liquids technology during the apartheid era. Build enough coal-to-liquids plants and places like Montana and West Virginia could one day become major petroleum producers.

In Fort McMurray, what might be called the world's first unconventional oil boom is already under way. Since 2002, Shell, ConocoPhillips, Chevron, and Imperial Oil, which is primarily owned by ExxonMobil, have all received approval to construct major projects in the tar sands; Total has announced its intention to follow suit. Over the next five years, investment in the Fort McMurray area is expected to amount to more than seventy-five billion dollars. Residents of the town have taken to calling it Fort McMoney.

Thanks in large part to what's happening in the tar sands—output now tops a million barrels a day—Canada has become America's number one source of imported oil; the country supplies the United States with more petroleum than all of the nations of the Persian Gulf combined. (If you have bought gas recently in Colorado, Ohio, or Indiana—states where tar-sands oil is refined—you are probably driving around with a piece of northern Alberta in your tank.) By 2010, the tar sands' yield is expected to double, and by 2015, to triple. Crude from the tar sands and other unconventional sources could keep oil flowing well into the middle of the century, and perhaps beyond. Depending on how you look at things, this is either a heartening prospect or a terrifying one.

The company that has been producing oil from the tar sands the longest is known as Suncor. (Suncor used to be a part of Sun Oil, now Sunoco, but today it is owned and operated independently.) One day in the summer of 2007, I went to take a tour of its operations, which sprawl across several hundred square miles. I was picked up at the entrance to the site by a grandmotherly guide named Gloria Jackson, and together we went to fetch another Suncor official, named Darin Zandee. "There's no blasts today, so that's good," Zandee said, referring to the charges that are periodically set off to loosen the sands. We drove up to a lookout, from which we could see, spread before us, Suncor's newest mine, the Millennium. Rings of jet-black earthworks were scattered across an enormous pit, an arrangement that might have been based on a blueprint from the *Inferno*.

The Millennium Mine opened in 2002. Suncor expects to continue to pull tar sands out of it for the next twenty-five years. By then the pit, which is now roughly two miles in diameter, will be six miles across. We drove over the edge of the mine and slowly made our way down to the bottom. There a huge, Mike Mulligan-esque shovel was standing idle. Its bucket hung in midair, steel teeth glinting. Zandee said that to lift one of the teeth would require thirty men—"That gives you a sense of the scale." A gargantuan truck rumbled by. Zandee estimated that it was carrying about three hundred tons. "That's some of our smaller equipment," he said. The largest truck in the mine—the Caterpillar 797B—can haul more than four hundred tons. It has twelve-foot-tall tires, and its cab sits twenty-one feet off the ground. Driving one, I was told, is like trying to steer a house while peering out the window of the upstairs bathroom.

At the Millennium, the tar sands start at a depth of roughly a hundred feet and extend down in a more or less continuous layer, known as the "feed," for about a hundred and fifty feet. Before mining begins, everything above the feed—trees, bushes, grass, soil, rocks, wildlife— gets scooped up and carted away. (The material is delicately referred to as "overburden.") Below the tar sands, there's a thick layer of limestone, the remains of an ancient ocean that once covered Alberta. Suncor mines some of the limestone, too, and uses it to shore up the roads in the pit. What with the overburden and the tar sands and the limestone, Zandee said, "We try to move a million tons a day." He pointed out a truck in the distance that was dumping a load of tar sands onto what looked like a large platform. The platform was actually a grate, through which the sands were being fed into a giant tank of hot water.

In any given load of sands, only about 10 percent is

bitumen; to produce synthetic crude, the other 90 percent has to be separated out. In the hot-water tank, the sands get spun around; the liberated bitumen is then siphoned off. For every barrel of synthetic crude that Suncor eventually produces, forty-five hundred pounds of tar sands have to be dug up and separated.

We made our way out of the pit and headed on, following the bitumen to its next stop, the upgrader. Along the way, we passed a murky expanse of water with oily scum on the surface. A few dozen scarecrow-like creatures, fixed to empty barrels, were bobbing on top. This, Gloria Jackson explained, was a tailings pond; it held water that had been used in the separation process and was too contaminated with mercury and other toxins to be released back into the Athabasca. (Suncor has nine such ponds, which collectively cover an area of eleven square miles.) The scarecrows, known as "bitu-men," were supposed to discourage birds from landing on the pond and poisoning themselves. Every minute or so, a dull boom filled the air. This was the sound of a propane cannon, another bird-intimidation device.

The primary difference between bitumen and ordinary crude is the size of the hydrocarbon molecules: in liquid oil, these molecules contain between five and twenty carbon atoms, while in bitumen they contain more than twenty. (At room temperature, pure bitumen is so viscous that it will not flow.) The main job of the upgrader is to break down the oversized hydrocarbons into smaller units. We drove along roads with names like Sulphur Street and

Diesel Alley and pulled up to a huge refinery-like complex that covered several square blocks. There were dozens of smokestacks and tanks, and more pipes than could possibly be counted. Jackson explained that somewhere inside this maze the bitumen would be "cracked," at a temperature of nearly nine hundred degrees. After that, in the form of synthetic crude, it would be piped to specially outfitted refineries, either in the United States or Canada, to be converted largely into transportation fuels—gasoline for cars, diesel for trucks, and jet fuel for planes. (Suncor owns a refinery near Denver that processes tar-sands oil.) I had told Jackson that I had twin boys at home, and at the end of the tour she handed me two yellow Matchbox-size versions of the 797B.

American accounts usually give the start of the oil age as 1859, the year that a former railroad conductor named Edwin L. Drake drilled his first successful well, near Titusville, Pennsylvania. Canadian accounts go back a year earlier, to 1858, when a businessman named James Miller Williams decided to dig a well for drinking water outside the town of Bear Creek, Ontario. Instead of water, he struck oil.

Efforts to extract oil from the tar sands soon followed. Entrepreneurs and con men sunk dozens of wells around Fort McMurray in the second half of the nineteenth century. (One enterprising German immigrant who claimed to have struck oil apparently poured the stuff down the hole himself.) Eventually, it became clear that there was no oil, and attention turned to mining the bitumen. In 1930, a former farmer named Robert Fitzsimmons set up the first commercial separation plant in the tar sands; in 1938, Fitzsimmons had to flee Canada to avoid his creditors.

In 1956, an American geologist, Manley Natland, came up with the idea of streamlining the process by using atom bombs. Natland reasoned that "thermal devices" could be lowered into the limestone beneath the tar sands and exploded. This would create cavities into which the bitumen, heated to more than a thousand degrees, would flow and from which it could then be collected. The idea was taken seriously at the highest levels in both Ottawa and Washington—the United States Atomic Energy Commission even agreed to supply a bomb to test Natland's theory—but it was never implemented. (Beginning in the mid-1960s, the Soviet Union actually tried the experiment, setting off half a dozen nuclear explosions to stimulate conventional oil production; production increased, but, unfortunately, much of the oil turned out to be radioactive.)

The technology for removing bitumen from the tar sands is probably still best described as a work in progress. Where the feed lies closest to the surface, as, for example, at the Suncor site, the bitumen is strip-mined and then separated. But most of the tar sands lie too deep to be mined profitably. In these zones, a method known as in-situ extraction is used. In-situ extraction is based on much the same principle as Natland's scheme, minus the atom bombs. Typically, two horizontal wells are drilled into the sands, one above the other. High-pressure steam is injected into the top well; eventually, the tar sands grow hot enough—nearly four hundred degrees—that bitumen begins to flow into the bottom well. The technical name for this process is Steam Assisted Gravity Drainage, or SAGD (pronounced "sag-dee").

Whichever method is used, a great deal of energy is required. To produce a barrel of synthetic crude through mining takes roughly eight hundred and ten megajoules, which is the energy content of about an eighth of a barrel of oil. To produce a barrel of synthetic crude through SAGD takes more than sixteen hundred megajoules, which is the energy content of more than a quarter of a barrel of oil. This means that, for every three barrels extracted via SAGD, one has, in effect, been consumed.

Tar-sands oil itself could, in principle, be used to power the operations; in fact, most of the energy used to generate the steam for SAGD, as well as to run all the upgraders and separators, now comes from natural gas. It is estimated that by 2012 tar-sands operations will consume two billion cubic feet of natural gas a day, or enough to heat all the homes in Canada. Such is the demand for natural gas around Fort McMurray that a consortium of companies, including Shell Canada and Imperial Oil, has proposed building a 750-mile pipeline from the Arctic Ocean through the largely undisturbed wilderness of the Mackenzie River Valley and down into northern Alberta. The proposal, which has been challenged by native and environmental groups, has yet to receive regulatory approval; meanwhile, a variety of other plans have been floated. As it happens, while I was visiting Fort McMurray a company called the Energy Alberta Corporation filed an application to build a pair of nuclear reactors four hundred miles west of town. Early reports stated that the company already had a "large industrial off-taker" lined up to buy nearly three quarters of the twenty-two hundred megawatts that the reactors would generate. Energy Alberta would not disclose the identity of this "off-taker"; in the local press it seemed to be taken for granted that the power would be going to the tar sands.

There are several reasons that companies like Chevron and ExxonMobil are now rushing to develop the tar sands, the most obvious being that it's increasingly profitable to do so. Converting the sands into synthetic crude costs around thirty dollars a barrel; a few weeks after I visited the Millennium Mine, the price of a barrel of oil on the New York Mercantile Exchange climbed to more than ninety dollars. Other synthetic fuels require more elaborate processing and are commensurately more costly to produce; converting coal into oil, for example, requires gasifying the coal under intense pressure and heat, then condensing it into a liquid. Extracting oil from shale, meanwhile, involves basically rewriting geological history. (Shell has been experimenting with a process that involves baking the shale with electric heaters until it reaches a temperature of nearly seven hundred degrees while, at the same time, freezing the area around it.) If the price of oil remains above ninety dollars, then these and other un-conventional forms of fuel can also be developed at a profit, and, all other things being equal, they will be.

No matter how it is carried out, oil extraction is a destructive business. Conventional oil wells require pipe-lines and drill pads and roads for heavy equipment; all of these fragment (or destroy) the landscape. The flaring of natural gas, which often accompanies oil production, produces an array of air pollutants, and leaks and spills release toxins ranging from volatile chemicals, like benzene (a known carcinogen), to much heavier compounds, like benzopyrene (another known carcinogen). With uncon-ventional oil, the damage tends to be higher all around—more land gets disturbed, more pollutants are produced, and more opportunities arise for contamination. And then there are the greenhouse gases.

Alex Farrell is a professor in the Energy and Resources Group at the University of California at Berkeley who studies the impacts of unconventional oil. About a decade ago, Farrell realized that all the major climate models were based on the same faulty premise: they assumed that in the future increased oil demand would be met with increased supplies of conventional crude. Together with a graduate student named Adam Brandt, Farrell decided to try to come up with projections that more accurately reflected reality. For their calculations, the two assumed that where there was a gap between demand and conventional supply it would be filled with synthetic fuels, first with tar-sands oil and later with oil from coal and shale. (According to high-end estimates, coal and oil shale could together yield some ten trillion barrels of unconventional crude.) They then calculated what the impact would be on global carbon dioxide levels.

"All unconventional forms of oil are worse for green-house gas emissions than petroleum," Farrell told me. "And it's pretty easy to understand why. It's not so hard to turn liquid petroleum into liquid fuels. Turning a solid material like coal into a liquid—it sounds hard to do, and it is hard to do. And that extra effort shows up in higher energy consumption and higher water use and higher emissions." In the case of tar-sands oil, total greenhouse gas emissions per barrel—which is to say, the carbon dioxide produced in creating the oil and then burning it—are between 15 and 40 percent higher than those from conventional oil. In the case of coal-to-liquids, or CTL, total emissions are almost two times as high as with conventional oil, and for oil shale they can be more than twice as high.

"Let's take coal-to-liquids," Farrell said. "You're talking about nearly doubling the greenhouse gas emissions. Think about this—we're talking about a world in which overall greenhouse gas emissions should start to go down, and this is a technology that doubles emissions. They don't go together too well, do they?" Farrell and Brandt found that the shift to unconventional oil could add somewhere between fifty and four hundred gigatons of carbon to the atmosphere by 2100.

"The environment and climate change are what are called 'externalities,'" Farrell continued. "And at the moment we don't have effective ways of including these externalities in market transactions of any sort. Until we do, the market won't solve them, since by definition they're external to the market. They're a social good—government has to step up and say, 'We're going to take this into account.'"

One way that a government could take greenhouse gas emissions into account would be to tax them. This would encourage producers of unconventional fuels to cut their emissions, by, for example, employing "carbon capture and storage" technologies. Ideally, it would also prompt entrepreneurs to develop alternatives to oil, like biofuels. Many analyses, though, suggest that, to have an appreciable effect on the oil sector, carbon taxes would have to be quite high—in the neighborhood of two dollars on a gallon of gasoline—precisely because today there are no readily available substitutes for gas or diesel or jet fuel. Farrell favors federal fuel standards, which would function somewhat like vehicle-efficiency standards, requiring oil companies to achieve a certain emissions target across all the products that they sell. (This target could be adjusted over time, much as auto-efficiency standards were ratcheted up during the seventies and eighties.) California has drawn up such a plan—the California Low Carbon Fuel Standard—and several bills have been introduced in Congress that would impose such standards nationally.

At the same time, there is a good deal of support in

Washington for measures that would, in effect, subsidize high-carbon fuels. One such measure, the Coal-to-Liquid Fuel Promotion Act, introduced in 2007 by then-senators Jim Bunning, of Kentucky, and Barack Obama, of Illinois, would have encouraged companies to invest in CTL plants by providing tax incentives and federal loan guarantees. (That bill never got out of committee.) Although CTL would be profitable at today's oil prices, building the plants requires large capital investments, which are considered risky as long as there's a chance that oil prices will fall.

"If companies could lay off the risk of oil prices dropping below forty dollars a barrel, there would be enormous investment in this," Farrell told me. "But, when policies are proposed to promote CTL, I think the question to ask is, Is this an industry we want to start now?"

The Athabasca River flows north, into Lake Athabasca, which spans the Alberta-Saskatchewan border. In the winter, it is possible to drive the 150 miles from Fort McMurray to the lake on an ice road. (Because of rising temperatures, the number of days that the road is passable has been steadily shrinking.) In the summer, the only way to make the trip is by boat or by prop plane. One day when I was visiting Alberta, I flew up to a village on the edge of the lake, Fort Chipewyan, in a six-seat Cessna. As the plane gained altitude, I could see the vast black pits of the tar-sands mines that surround Fort McMurray. Farther north, the pits gave way to regularly spaced square-shaped clearings in the trees—signs of preparation for in-situ operations. Finally, these, too, gave way, and below was nothing but the wild green of the boreal forest. (Spread over 1.4 billion acres, Canada's boreal forest is considered one of the largest still intact ecosystems on the planet.)

Fort Chipewyan, which was founded in the 1780s as a trading post, is a native village; about half its twelve hundred or so residents are Mikisew Cree, and the other half are Athabasca Chipewyan. It has a few hundred houses, a post office, and two churches—one Anglican and one Catholic—both perched near the edge of the lake. To a certain extent, Fort Chip, as it is known locally, has shared in the tar-sands boom; many residents of the village work construction jobs in Fort McMurray and return home only on their days off. At the same time, there's a good deal of concern in the village about what is happening. A peculiarly high number of cases of a rare cancer have been reported in town; this has prompted speculation that toxins from the tailings ponds are working their way downriver into the lake, which provides the village with drinking water as well as with staples like whitefish and pike. Meanwhile, both the Chipewyan and the Cree consider many of the tracts that the Alberta government has leased to oil companies to be their ancestral lands. The week before I visited Fort Chip, there was a rally at the local community center, calling for a moratorium on new projects.

"It's sad to see this thing destroyed, you know," Ray Ladouceur, a fisherman I met, said. We were standing by the lake, which is more than two hundred miles long. It was a still afternoon, and billowy white clouds were reflected in the water. "A lot of the fish are getting—I might as well say it—scabby.

"I don't know what we have to do to try to prevent them from destroying any more," he said, referring to the oil companies. "They try to say they can clean it. There's no way. It'll take a thousand years before it flushes itself out, and I think I'll be too damn old for that."

In recent years, opposition to new tar-sands projects has been steadily growing. Around Fort McMurray, the emphasis is on local impacts; town officials have fought recent expansion proposals by several oil companies on the ground that there's already a shortage of housing and hospital beds in the area. In the rest of Canada, the focus is on the destruction of the boreal forest and the implications for the climate. Canada, in contrast to the United States, was an early signatory to the Kyoto Protocol, but it will be all but impossible for the country to meet its CO_2-reduction goals, in part because of the tar sands. (A Toronto *Globe and Mail* op-ed piece on emissions from the sands that ran not long before I visited Fort McMurray was titled "The Gassy Elephant in Our Living Room.") The former Canadian environment minister Charles Caccia has compared the country's position on greenhouse gases—pledging to reduce emissions on the one hand while increasing tar-sands production on the other—to "attempting to ride two horses galloping in opposite directions."

Meanwhile, development in northern Alberta continues

unabated. All the applications opposed by Fort McMurray officials were ultimately approved, and just a few months ago an American company, Hyperion Resources, announced plans to build the first new oil refinery in this country in thirty years, to handle increasing volumes of tar-sands crude. Stéphane Dion, the leader of Canada's Liberal Party (which is currently out of power), has said, "There is no environmental minister on earth who can stop the oil from coming out of the sand, because the money is too big."

When I first landed at Fort Chip's tiny airport, the place was deserted. When I returned there for the flight back, I found a few dozen people standing on the tarmac. The crowd, I was told, was waiting for a corpse; a village elder had died the previous day in a hospital in Fort McMurray, and his body was being brought home. Everyone was quiet as the casket was carried out of the plane and then loaded onto the back of a pickup truck. As soon as the crowd dispersed, I and three other passengers climbed into the Cessna, and two minutes later we took off. Below was the wilderness, then the perfectly square clearings in the trees, and, finally, as we headed into Fort McMurray, the vast pits and the black ponds, with the bitu-men bobbing on top.

Originally published November 2007

Chapter 14

THE ISLAND IN THE WIND

JØRGEN TRANBERG IS A farmer who lives on the Danish island of Samsø. He is a beefy man with a mop of brown hair and an unpredictable sense of humor. When I arrived at his house, one gray morning in spring, he was sitting in his kitchen, smoking a cigarette and watching grainy images on a black-and-white TV. The images turned out to be closed-circuit shots from his barn. One of his cows, he told me, was about to give birth, and he was keeping an eye on her. We talked for a few minutes, and then, laughing, he asked me if I wanted to climb his wind turbine. I was pretty sure I didn't, but I said yes anyway.

We got into Tranberg's car and bounced along a rutted dirt road. The turbine loomed up in front of us. When we reached it, Tranberg stubbed out his cigarette and opened a small door in the base of the tower. Inside were eight ladders, each about twenty feet tall, attached one above the other. We started up and were soon huffing. Above the last ladder, there was a trapdoor, which led to a sort of engine room. We scrambled into it, at which point we were standing on top of the generator. Tranberg pressed a button, and the roof slid open to reveal the gray sky and a patchwork of green and brown fields stretching toward the sea. He pressed another button. The rotors, which he had switched off during our climb, started to turn, at first sluggishly and then much more rapidly. It felt as if we were about to take off. I'd like to say the feeling was exhilarating; in fact, I found it sickening. Tranberg looked at me and started to laugh.

Samsø, which is roughly the size of Nantucket, sits in what's known as the Kattegat, an arm of the North Sea. The island is bulgy in the south and narrows to a bladelike point in the north so that on a map it looks a bit like a woman's torso and a bit like a meat cleaver. It has twenty-two villages that hug the narrow streets; out back are fields where farmers grow potatoes and wheat and strawberries. Thanks to Denmark's peculiar geography, Samsø is smack in the center of the country and, at the same time, in the middle of nowhere.

For the past decade or so, Samsø has been the site of an unlikely social movement. When it began, in the late 1990s, the island's forty-three hundred inhabitants had what might be described as a conventional attitude toward energy: as long as it continued to arrive, they weren't much interested in it. Most Samsingers heated their houses with oil, which was brought in on tankers. They used electricity imported from the mainland via cable, much of which was generated by burning coal. As a result, each Samsinger put into the atmosphere, on average, nearly eleven tons of carbon dioxide annually.

Then, quite deliberately, the residents of the island set about changing this. They formed energy cooperatives and organized seminars on wind power. They removed their furnaces and replaced them with heat pumps. By 2001, fossil fuel use on Samsø had been cut in half. By 2003, instead of importing electricity, the island was exporting it, and by 2005, it was producing from renewable sources more energy than it was using.

The residents of Samsø that I spoke to were clearly proud of their accomplishment. All the same, they insisted on their ordinariness. They were, they noted, not wealthy, nor were they especially well educated or idealistic. They weren't even terribly adventuresome. "We are a conservative farming community" is how one Samsinger put it. "We are only normal people," Tranberg told me. "We are not some special people."

This year, the world is expected to burn through some thirty-one billion barrels of oil, six billion tons of coal, and a hundred trillion cubic feet of natural gas. The combustion of these fossil fuels will produce, in aggregate, some four hundred quadrillion BTUs of energy. It will also yield around eight billion tons of carbon. Next year, global consumption of fossil fuels is expected to grow by about 2 percent, meaning that emissions will rise by more than a hundred million tons, and the following year consumption is expected to grow by yet another 2 percent.

When carbon dioxide is released into the air, about a

third ends up, in relatively short order, in the oceans. (CO_2 dissolves in water to form a weak acid; this is the cause of the phenomenon known as "ocean acidification.") A quarter is absorbed by terrestrial ecosystems—no one is quite sure exactly how or where—and the rest remains in the atmosphere. If current trends in emissions continue, then sometime within the next four or five decades the chemistry of the oceans will have been altered to such a degree that many marine organisms—including reef-building corals—will be pushed toward extinction. Meanwhile, atmospheric CO_2 levels are projected to reach 550 parts per million—twice preindustrial levels—virtually guaranteeing an eventual global temperature increase of three or more degrees.

Today, with CO_2 levels at 390 parts per million, the disruptive impacts of climate change are already apparent. The Arctic ice cap, which has shrunk by half since the 1950s, is melting at an annual rate of twenty-four thousand square miles, meaning that an expanse of ice the size of West Virginia is disappearing each year. Over the past ten years, forests covering a hundred and fifty million acres in the United States and Canada have died from warming-related beetle infestations. It is believed that rising temperatures are contributing to the growing number of international refugees—"Climate change is today one of the main drivers of forced displacement," the United Nations' high commissioner for refugees, António Guterres, said recently—and to armed conflict: some experts see a link between the fighting in Darfur, which has claimed as many as three hundred thousand lives, and changes in rainfall patterns in equatorial Africa.

"If we keep going down this path, the Darfur crisis will be only one crisis among dozens of others," then-president Nicolas Sarkozy, of France, told a meeting of world leaders in April 2008. The secretary-general of the United Nations, Ban Ki-moon, has called climate change "the defining challenge of our age."

In the context of this challenge, Samsø's accomplishments could be seen as trivial. Certainly, in numerical terms they don't amount to much: all the island's avoided emissions of the past ten years are overwhelmed by the CO_2 that a single coal-fired power plant will emit in the next three weeks, and China is building new coal-fired plants at the rate of roughly four a month. But it is also in this context that the island's efforts are most significant. Samsø transformed its energy systems in a single decade. Its experience suggests how the carbon problem, as huge as it is, could be dealt with, if we were willing to try.

Samsø set out to reinvent itself thanks to a series of decisions that it had relatively little to do with. The first was made by the Danish Ministry of Environment and Energy in 1997. The ministry, looking for ways to promote innovation, decided to sponsor a renewable-energy contest. In order to enter, a community had to submit a plan showing how it could wean itself off fossil fuels. An engineer who didn't actually live on Samsø thought the island would make a good candidate. In consultation with

Samsø's mayor, he drew up a plan and submitted it. When it was announced that Samsø had won, the general reaction among residents was puzzlement. "I had to listen twice before I believed it," one farmer told me.

The brief surge of interest that followed the announcement soon dissipated. Besides its designation as Denmark's "renewable-energy island," Samsø received basically nothing—no prize money or special tax breaks, or even government assistance. One of the few people on the island to think the project was worth pursuing was Søren Hermansen.

Hermansen is a compact man with close-cropped hair, ruddy cheeks, and dark blue eyes. He was born on Samsø and, save for a few stints away, to travel and go to university, has lived there his entire life. His father was a farmer who grew, among other things, beets and parsley. Hermansen, too, tried his hand at farming—he took over the family's hundred acres when his father retired—but he discovered he wasn't suited to it. "I like to talk, and vegetables don't respond," he told me. He leased his fields to a neighbor and got a job teaching environmental studies at a local boarding school. Hermansen found the renewable-energy-island concept intriguing. When some federal money was found to fund a single staff position, he became the project's first employee.

For months, which stretched into years, not much happened. "There was this conservative hesitating, waiting for the neighbor to do the move," Hermansen recalled. "I know the community and I know this is what usually happens." Rather than working against the islanders' tendency to look to one another, Hermansen tried to work with it.

"One reason to live here can be social relations," he said. "This renewable-energy project could be a new kind of social relation, and we used that." Whenever there was a meeting to discuss a local issue—any local issue— Hermansen attended and made his pitch. He asked Samsingers to think about what it would be like to work together on something they could all be proud of. Occasionally, he brought free beer along to the discussions. Meanwhile, he began trying to enlist the support of the island's opinion leaders. "This is where the hard work starts, convincing the first movers to be active," he said. Eventually, much as Hermansen had hoped, the social dynamic that had stalled the project began to work in its favor. As more people got involved, that prompted others to do so. After a while, enough Samsingers were participating that participation became the norm.

"People on Samsø started thinking about energy," Ingvar Jørgensen, a farmer who heats his house with solar hot water and a straw-burning furnace, told me. "It became a kind of sport."

"It's exciting to be a part of this," Brian Kjær, an electrician who installed a small-scale turbine in his backyard, said. Kjær's turbine, which is seventy-two feet tall, generates more current than his family of three can use, and also more than the power lines leading away from his house can handle, so he uses the excess to heat water, which he stores in a tank that he rigged up in his garage. He told me that one day he would like to use the leftover electricity to produce hydrogen, which could potentially run a fuel-cell car.

"Søren, he has talked again and again, and slowly it's spread to a lot of people," he said.

Since becoming the "renewable-energy island," Samsø has increasingly found itself an object of study. Researchers often travel great distances to get there, a fact that is not without its own irony. The day after I arrived, from New York via Copenhagen, a group of professors from the University of Toyama, in Japan, came to look around. They had arranged a tour with Hermansen, and he invited me to tag along. We headed off to meet the group in his electric Citroën, which is painted blue with white puffy clouds on the doors. It was a drizzly day, and when we got to the dock the water was choppy. Hermansen commiserated with the Japanese, who had just disembarked from the swaying ferry; then we all boarded a bus.

Our first stop was a hillside with a panoramic view of the island. Several wind turbines exactly like the one I had climbed with Tranberg were whooshing nearby. In the wet and the gray, they were the only things stirring. Off in the distance, the silent fields gave way to the Kattegat, where another group of turbines could be seen, arranged in a soldierly line in the water.

All told, Samsø has eleven large land-based turbines. (It has about a dozen additional micro-turbines.) This is a lot of turbines for a relatively small number of people, and the ratio is critical to Samsø's success, as is the fact that the wind off the Kattegat blows pretty much continuously; flags on Samsø, I noticed, do not wave—they stick straight out, as in children's drawings. Hermansen told us that the land-based turbines are a hundred and fifty feet tall, with rotors that are eighty feet long. Together, they produce some twenty-six million kilowatt-hours a year, which is just about enough to meet all the island's demands for electricity. (This is true in an arithmetic sense; as a practical matter, Samsø's production of electricity and its needs fluctuate, so that sometimes it is feeding power into the grid and sometimes it is drawing power from it.) The offshore turbines, meanwhile, are even taller—a hundred and ninety-five feet high, with rotors that extend a hundred and twenty feet. A single offshore turbine generates roughly eight million kilowatt-hours of electricity a year, which, at Danish rates of energy use, is enough to satisfy the needs of some two thousand homes. The offshore turbines—there are ten of them—were erected to compensate for Samsø's continuing use of fossil fuels in its cars, trucks, and ferries. Their combined output, of around eighty million kilowatt-hours a year, provides the energy equivalent of all the gasoline and diesel oil consumed on the island, and then some; in aggregate, Samsø generates about 10 percent more power than it consumes.

"When we started, in 1997, nobody expected this to happen," Hermansen told the group. "When we talked to local people, they said, Yes, come on, maybe in your dreams." Each land-based turbine cost the equivalent of eight hundred and fifty thousand dollars. Each offshore turbine cost around three million dollars. Some of Samsø's turbines were erected by a single investor, like Tranberg; others were purchased collectively. At least four hundred and fifty island residents own shares in the onshore turbines, and a roughly equal number own shares in those offshore. Shareholders, who also include many nonresidents, receive annual dividend checks based on the prevailing price of electricity and how much their turbine has generated.

"If I'm reduced to being a customer, then if I like something I buy it, and if I don't like it I don't buy it," Hermansen said. "But I don't care about the production. We care about the production, because we own the wind turbines. Every time they turn around, it means money in the bank. And, being part of it, we also feel responsible." Thanks to a policy put in place by Denmark's government in the late 1990s, utilities are required to offer ten-year fixed-rate contracts for wind power that they can sell to customers elsewhere. Under the terms of these contracts, a turbine should—barring mishap—repay a shareholder's initial investment in about eight years.

From the hillside, we headed to the town of Ballen. There we stopped at a red shed-shaped building made out of corrugated metal. Inside, enormous bales of straw were stacked against the walls. Hermansen explained that the building was a district heating plant that had been designed to run on biomass. The bales, each representing the equivalent of fifty gallons of oil, would be fed into a furnace, where water would be heated to a 138 degrees. This hot water would then be piped underground to 260 houses in Ballen and in the neighboring town of Brundby. In this way, the energy of the straw burned at the plant would be transferred to the homes, where it could be used to provide heat and hot water.

Samsø has two other district heating plants that burn straw—one in Tranebjerg, the other in Onsbjerg—and also a district plant, in Nordby, that burns wood chips. When we visited the Nordby plant, later that afternoon, it was filled with what looked like mulch. (The place smelled like a potting shed.) Out back was a field covered in rows of solar panels, which provide additional hot water when the sun is shining. Between the rows, sheep with long black faces were munching on the grass. The Japanese researchers pulled out their cameras as the sheep snuffled toward them, expectantly.

Of course, burning straw or wood, like burning fossil fuels, produces CO_2. The key distinction is that while fossil fuels release carbon that otherwise would have remained sequestered, biomass releases carbon that would have entered the atmosphere anyway, through decomposition. As long as biomass regrows, the CO_2 released in its combustion should be reabsorbed, meaning that the cycle is—or at least can be—carbon neutral. The wood chips used in the Nordby plant come from fallen trees that previously would have been left to rot. The straw for the Ballen-Brundby plant comes mainly from wheat stalks that would previously have been burned in the fields. Together, the biomass heating plants prevent the release of some twenty-seven hundred tons of carbon dioxide a year.

In addition to biomass, Samsø is experimenting on a modest scale with biofuels: a handful of farmers have converted their cars and tractors to run on canola oil. We stopped to visit one such farmer, who grows his own seeds, presses his own oil, and feeds the leftover mash to his cows. The farmer couldn't be located, so Hermansen started up the press himself. He stuck a finger under the spout, then popped it into his mouth. "The oil is very good," he announced. "You can use it in your car, and you can use it on your salad."

After the tour, I went back with Hermansen to his office, in a building known as the Energiakademi. The academy, which looks like a Bauhaus interpretation of a barn, is covered with photovoltaic cells and insulated with shredded newspapers. It is supposed to serve as a sort of interpretive center, though when I visited, the place was so new that the rooms were mostly empty. Some high school students were kneeling on the floor, trying to put together a miniature turbine.

I asked Hermansen whether there were any projects that hadn't worked out. He listed several, including a plan to use natural gas produced from cow manure and an experiment with electric cars that failed when one of the demonstration vehicles spent most of the year in the shop. The biggest disappointment, though, had to do with consumption.

"We made several programs for energy savings," he told me. "But people are acting—what do you call it?—irresponsibly. They behave like monkeys." For example, families that insulated their homes better also tended to heat more rooms, "so we ended up with zero." Essentially, he said, energy use on the island has remained constant for the past decade.

I asked why he thought the renewable-energy-island effort had got as far as it had. He said he wasn't sure, because different people had had different motives for participating. "From the very egoistic to the more overall perspective, I think we had all kinds of reasons."

Finally, I asked what he thought other communities might take from Samsø's experience.

"We always hear that we should think globally and act locally," he said. "I understand what that means—I think we as a nation should be part of the global consciousness. But each individual cannot be part of that. So 'Think locally, act locally' is the key message for us."

"There's this wish for showcases," he added. "When we are selected to be the showcase for Denmark, I feel ashamed that Denmark doesn't produce anything bigger than that. But I feel proud because we are the showcase. So I did my job, and my colleagues did their job, and so did the people of Samsø."

Around the same time that Samsø was designated Denmark's renewable-energy island, a group of Swiss scientists who were working on similar issues performed a thought experiment. The scientists, all of whom were affiliated with the Swiss Federal Institute of Technology, asked themselves what level of energy use would be sustainable, not just for an island or a small European nation but for the entire world. The answer they came up with—two thousand watts per person—furnished the name for a new project: the 2,000-Watt Society.

"What's important, I think, to know is that the 2,000-Watt Society is not a program of hard life," the director of the project, Roland Stulz, told me when I went to speak to him at his offce, in the Zurich suburb of Dübendorf. "It is not what we call *Gürtel enger schnallen*"—belt tightening— "it's not starving, it's not having less comfort or fun. It's a creative approach to the future."

Stulz, who is sixty-three, is a soft-spoken man with dark wavy hair and a salt-and-pepper mustache. He was trained as an architect and later became interested in energy-efficient building. In 2001, when he took over the 2,000-Watt Society, his mandate was to push it into the realm of the practical. (His work is funded in part by the Swiss Federal Institute of Technology, which has campuses in Zurich and Lausanne, and in part by private donations.) He began holding meetings that brought researchers together with government officials from cities like Zurich and Basel.

"I divided them into groups," Stulz recalled. "And I told them, 'At four o'clock each group must come and tell the whole session what project they will do in the future, and who will lead the projects.' And they said, 'Oh, it's not possible.' But at four o'clock everybody came with a project. And that's how we started." The cantons of Geneva and Basel-Stadt and the city of Zurich subsequently endorsed the aims of the 2,000-Watt Society, as did the Swiss Federal Department of the Environment, Transport, Energy, and Communications. "At first glance, the objective of a two-thousand-watt society appears unrealistic," Moritz Leuenberger, the head of the federal department, has said. "But the necessary technology already exists."

One afternoon, Stulz took me to visit the headquarters of an aquatic-research center known as Eawag, which was designed to meet the 2,000-Watt Society's energy-efficiency goals. (Eawag is an acronym for a German name so complicated that even German speakers can't remember it.) We drove over in his Volvo, which runs on compressed natural gas produced in part from rotting vegetables. When I first caught sight of the place, I thought it was covered with banners; these turned out to be tinted-glass panels. Inside, hanging from a set of chains in a large atrium, was what I took to be a sculpture of a bug. This turned out to be a model of a water molecule, enlarged some ten billion times.

Among the many unusual features of the Eawag Center is a lack of usual features. The building, which opened in 2006, has no furnace; it is so tightly insulated that, on most days, the warmth thrown off by the office equipment and the two hundred people who work inside is enough to keep it comfortable. Additional heat is provided by the sun—in winter, the outside panels tilt to allow in the maximum amount of light—and by air sucked in from underground. The building also has no conventional air conditioners: in summer, the panels tilt to provide shade, and if the building gets hot during the day, at night the windows at the top of the atrium open, and the warm air rushes out. It supplies about a third of its own electricity with photovoltaic panels installed on the roof and gets its hot water from solar collectors. Its bathrooms are equipped with specially designed "no mix" toilets that separate out urine, which contains potentially useful phosphorus and nitrogen. ("Exploiting common waste as a resource is a mark of sustainable civilization," a booklet on the building observes.)

"It's not a miracle, such a building," Stulz told me when we went to have a cup of coffee in the center's cheerfully modernist cafeteria. "It's just putting smart elements together in a smart way." Outside, it was rainy and forty-three degrees; inside the temperature was a pleasant seventy.

One way to think about the 2,000-Watt Society is in terms of lightbulbs. Let's say you turn on twenty lamps, each with a hundred-watt bulb. Together, the lamps will draw two thousand watts of power. Left on for a day, they will consume forty-eight kilowatt-hours of energy; left on for a year, they will consume 17,520 kilowatt-hours. A person living a two-thousand-watt life would consume in all his activities—working, eating, traveling—the same amount of energy as those twenty bulbs, or 17,520 kilowatt-hours annually.

Most of the people in the world today consume far less than this. The average Bangladeshi, for example, uses only about twenty-six hundred kilowatt-hours a year—this figure includes all forms of energy, from electricity to transportation fuel—which is the equivalent of using roughly three hundred watts continuously. The average Indian uses about eighty-seven hundred kilowatt-hours a year, making India a one-thousand-watt society, while the average Chinese uses about thirteen thousand kilowatt-hours a year, making China a fifteen-hundred-watt society.

Those of us who live in the industrialized world, by contrast, consume far more than two thousand watts. Switzerland, for instance, is a five-thousand-watt society. Most other Western European countries are six-thousand-watt societies; the United States and Canada run at twelve thousand watts. One of the founding principles of the 2,000-Watt Society is that this disparity is in itself unsustainable. "It's a basic matter of fairness" is how Stulz put it to me. But increasing energy use in developing countries to match that of industrialized nations would be unacceptable on ecological grounds. Were per-capita demand in the developing world to reach current European levels, global energy consumption would more than double, and were it to rise to the American level, global energy consumption would more than triple. The 2,000-Watt

Society gives industrialized countries a target for cutting energy use at the same time that it sets a limit for growth in developing nations.

The last time Switzerland was a two-thousand-watt society was in the early 1960s. By the end of that decade, energy use had reached three thousand watts, and by the mid-seventies it was up to four thousand watts. This rapid rise could be said to follow from technological advances—the spread of automobiles, the advent of jet travel, the proliferation of appliances and electronic devices—or it could be seen as just the reverse: a failure to apply technology where it is needed. A few years ago, a group of Swiss scientists published a white paper—or, to use the Swiss term, a "white book"—on the feasibility of a 2,000-Watt Society. Relying on widely agreed-upon figures, the scientists estimated that two thirds of all the primary energy consumed in the world today is wasted, mostly in the form of heat that nobody wants or uses. ("Primary energy" is the energy contained in, say, a lump of coal; "useful energy" is the light emitted by a bulb once that coal has been burned to produce steam, the steam has been used to run a turbine, and the resulting electricity has been transmitted over the grid to heat the bulb's filament.) This same paper concluded that, with currently available technologies, buildings could be made 80 percent more efficient, cars 50 percent more efficient, and motors 25 percent more efficient.

In Switzerland, I visited several other buildings that, like the Eawag Center, had been specifically designed to maximize efficiency. One was an upscale apartment building in Basel. The apartments have eighteen-inch-thick walls filled with insulation, triple-paned windows coated with a special reflective film, and a heat-recovery system that captures 80 percent of the energy normally lost through ventilation. Instead of a boiler, it has a geothermal heat pump, which essentially sucks energy out of the groundwater. In the summer, the same system is used for cooling. (In compliance with Swiss building codes, the building also contains a bomb shelter.)

"The construction industry is very traditional," Franco Fregnan, an engineer who showed me around the apartments, said. "If you bring an innovation to them, you usually have to wait another generation until it arrives into a building. And we are trying to change that, step by step."

"It usually makes sense to become more intelligent in any human activity," Stulz told me. "As the former Saudi Arabian oil minister Sheikh Yamani once said, the Stone Age didn't end because there were no more stones. It ended because people became more intelligent."

What would it take to lead a two-thousand-watt life? When I posed this question to Stulz, he gave me another research paper, which offers case studies of six fictionalized households. The Jeannerets are an imaginary family of four who live in Glattbrugg, a town north of Zurich. They own an energy-efficient house, travel by electric bike or train, and occasionally rent a car-they belong to a car—sharing service—to do their grocery shopping. The Moeris, fictional farmers who live northeast of Bern, generate their own electricity with natural gas produced from cow manure; and Alain, Michel, Angela, and Marlène, fictional students living in Geneva, share all their appliances, use the tram, and like to go hiking in the French Alps during school breaks. "There is no formula for how to achieve a two-thousand-watt society," the paper declares. "Three things are needed: societal decisions . . . technical innovation, and the resolve of every individual to act in an energy-conscious way."

Very broadly speaking, the average Swiss today uses energy as follows: 1,500 watts per day for living and office space (this includes heat and hot water), 1,100 watts for food and consumer items (the energy that it takes to produce and transport goods is referred to as "embodied" or "gray" energy), 600 watts for electricity, 500 watts for automobile travel, 250 watts for air travel, and 150 watts for public transportation. Each person's share of Switzerland's public infrastructure, which includes facilities like water- and sewage-treatment plants, comes to nine hundred watts. Reducing these five thousand watts to two thousand would seem to require a significant reduction in every realm. Assuming that infrastructure-related consumption could be cut to five hundred watts, a person who continued to use fifteen hundred watts for living and office space would have nothing left for food, electricity, and transportation. Similarly, a person who continued to travel and use electricity at current rates would consume two thousand watts without having anywhere to live or work, or anything to eat.

While I was in Switzerland, I kept looking for people who actually led two-thousand-watt lives.

"I'm pretty close, except for this stupid air travel," Gerhard Schmitt, the vice president for planning and logistics at the Zurich campus of the Swiss Federal Institute of Technology, told me. "I go once to Shanghai and it's gone." (A round-trip flight between Zurich and Shanghai is the equivalent of using something like eight hundred watts continuously for a year.)

"Let's skip that question," Stulz said when I put it to him. While he lives in an energy-efficient apartment, he, too, travels a great deal; when I visited, he had just returned from a conference in New Delhi, a round-trip that used roughly the equivalent of six hundred watts for the year.

The one person I spoke to who did seem to be leading a two-thousand-watt life, or something very near to it, was an engineer named Robert Uetz. Uetz works in the same building as Stulz, and when we returned from visiting the Eawag Center he was still in his office, even though it was after six. Stulz encouraged me to go talk to him.

"We don't experience it as a restriction," Uetz told me of his two-thousand-watt life style. "On the contrary. I don't feel that we're giving up anything." Uetz and his wife, a dentist, live with their two children in the city of Winterthur, near Zurich. About ten years ago, they bought a two-thousand-square-foot house in a newly built energy-efficient development. The house is heated with a geothermal heat pump—"It's crazy to heat a house with fossil fuels," Uetz said—and has a solar hot-water system. Uetz added photovoltaic panels to the roof to produce electricity; in the winter the panels produce somewhat less power than the house uses—it's equipped with the most energy-efficient lights and appliances the family could find—and in the summer they produce somewhat more, so that over the course of the year the house's electricity use nets out to zero.

"The most important decision was that we wouldn't have a car," Uetz told me. "That was a conscious decision. We looked for a house where we didn't need a car." Driving a lot—even in what, by today's standards at least, counts as an energy-efficient vehicle—also makes it difficult to live within two thousand watts. A person who drives a Toyota Prius ten thousand miles a year consumes roughly 225 gallons of gasoline. This is equivalent to consuming around eight thousand kilowatt-hours, or to using nearly a thousand watts on a continuous basis. (For a family of four, the same gasoline consumption would come to almost two hundred and fifty watts per person.)

"It's a matter of what you're used to, but I find taking the train a lot more pleasant than driving," Uetz went on. "On the train I can work and relax. If I took a car, I'd have to worry about parking and traffic, rain, snow, and a certain number of people who can't drive but are on the road anyway." When Uetz and his family go on vacation, they travel by rail. "The only thing I'd say that is sort of a restriction is the flying," he said. "Because, obviously, with the train where you can go is limited. We can't go to China, or if we did it would take a week."

"I don't make a religion out of it," he added. "I wouldn't do it if I didn't feel good about it—it's how I like to live."

By the 2,000-Watt Society's own reckoning, cutting consumption is just half—or, perhaps more accurately, a quarter—of what needs to be done. The project's ultimate goal is a world where people consume no more than two thousand watts apiece and where fifteen hundred of those watts come from carbon-free sources. In such a world, everyone would use energy sparingly, like Robert Uetz, and generate it renewably, like Jørgen Tranberg. In such a world, filled with windmills and net-zero houses, carbon emissions would fall sharply, and the concentration of CO_2 in the atmosphere would slowly level off. But how realistic is such a scenario?

Before I left Switzerland to fly back to New York (a trip equivalent to using roughly two hundred and fifty watts continuously for a year), I went to speak to the president of the research council of the Swiss National Science Foundation, Dieter Imboden. Imboden, who is sixty-four, is a compact man with an oval face and silvery hair. He received his training in theoretical solid-state physics, later became interested in environmental physics, and for several years chaired the Swiss Federal Institute of Technology's environmental-sciences department. In the late nineties, he served as the director of the 2,000-Watt Society. He said that as a scientist he could see no technical barriers to creating a two-thousand-watt world.

"We are putting our mental energy into the wrong basket," he told me. "Nothing has to be reinvented—for an engineer it's not even a challenge."

"The problems of the twenty-first century are a different kind of problem," he went on. "And I think our society will be measured according to the solution of this new kind of problem, which cannot be solved with the same recipe as the flight to the moon, or the Manhattan Project. It's a qualitative difference—a paradigm change in the role of science for our society."

He continued. "The difficult thing is what I call 'constructed Switzerland.' You in America could call it 'constructed United States'—the buildings and how they are built, but also where they are built and, even more important, the roads, the railroads, the lines for energy, for wastewater, and so on. It's not economically feasible to replace everything in one instant." But since infrastructure should in any case be replaced at the rate of roughly 2 percent a year, if the project is approached incrementally, it's a different task. Then, Imboden said, "it suddenly *is* feasible."

As of yet, no one has undertaken a rigorous analysis of the economics of a transition to two thousand watts. Researchers have tended, rather, to focus on the price of stabilizing carbon-dioxide levels in the atmosphere at a given concentration—either, say, 550 parts per million, which is double preindustrial levels, or 450 parts, which, many climate scientists say, is the very highest level advisable. Perhaps the most often-cited economic study is the Stern Review, commissioned by the British government and named for its lead author, Sir Nicholas Stern, formerly the chief economist for the World Bank. The Stern Review, published in October 2006, concluded that greenhouse gas levels could be stabilized below double preindustrial concentrations at a cost to global GDP of around 1 percent a year. (The Stern Review considered not just carbon dioxide but other greenhouse gases, like methane and nitrous oxide, as well.) An analysis released last year by the Swedish utility Vattenfall, with research assistance from the American consulting firm McKinsey & Company, reached a similar conclusion: it determined that many measures to reduce carbon emissions, like improving building insulation, would save money, while others, like installing wind turbines, would carry a price. The Vattenfall report estimates that "if all low-cost opportunities are addressed," CO_2 levels could be stabilized at 450 parts per million with an annual expenditure of six tenths of 1 percent of global GDP.

Though 1 percent of the global economy is clearly a lot of money, in the grand scheme of things it's also clearly manageable. It is about a ninth of what's currently spent on health care, a seventh of what's spent on oil, and half of what's spent on defense. (More than 40 percent of all the world's military expenditures are made by the United States.) Perhaps most pertinent, it's a far smaller figure than the cost of inaction. The Stern Review projects that if current emissions trends are allowed to continue, the eventual damage from climate change will "be equivalent to losing at least 5% of global GDP each year, now and forever," and that "if a wider range of risks and impacts is taken into account" that figure could "rise to 20% of GDP or more."

Twenty years ago, NASA's chief climate scientist, James Hansen, testified on Capitol Hill about the dangers of global warming. Just a few days ago, Hansen returned to the Hill to testify again. "Now, as then, frank assessment of scientific data yields conclusions that are shocking to the body politic," he said. "Now, as then, I can assert that these conclusions have a certainty exceeding ninety-nine percent. The difference is that now we have used up all slack in the schedule." Hansen went on to warn that there would be no practical way to prevent "disastrous" climate change unless the next president and Congress act quickly to curb emissions. Few parts of the United States may be as windy as Samsø, or as well organized as Switzerland, but just about everywhere there are possibilities for generating energy more inventively and using it more intelligently. Realizing these possibilities will require a great deal of effort. We may well decide not to make this effort. Such a choice to put off change, however, will merely drive us toward it.

Originally published July 2008

FURTHER READING

Over the last decade, many very good books have been written on climate change. For those interested in reading more on the subject, here are some suggestions:

Climate Change: Picturing the Science by Gavin Schmidt and Joshua Wolfe (New York: W. W. Norton, 2009).
Eaarth: Making a Life on a Tough New Planet by Bill McKibben (New York: Henry Holt, 2010).
Fraser's Penguins: A Journey to the Future in Antarctica by Fen Montaigne (New York: Henry Holt, 2010).
Heatstroke: Nature in an Age of Global Warming by Anthony D. Barnosky (Washington, D.C.: Island Press, 2009).
How to Cool the Planet: Geoengineering and the Audacious Quest to Fix Earth's Climate by Jeff Goodell (New York: Houghton Mifflin Harcourt, 2010).
Merchants of Doubt: How a Handful of Scientists Obscured the Truth on Issues from Tobacco Smoke to Global Warming by Naomi Oreskes and Erik M. Conway (New York: Bloomsbury Press, 2010).
Our Choice: A Plan to Solve the Climate Crisis by Al Gore (New York: Rodale Books, 2009).
Storms of My Grandchildren: The Truth About the Coming Climate Catastrophe and Our Last Chance to Save Humanity by James Hansen (New York: Bloomsbury Press, 2009).

The Global Warming Reader edited by Bill McKibben (New York: O/R Books, 2011).

The Hockey Stick and the Climate Wars: Dispatches from the Front Lines by Michael E. Mann (New York: Columbia University Press, 2009).

The Weather of the Future: Heat Waves, Extreme Storms, and Other Scenes from a Climate-Changed Planet by Heidi Cullen (New York: HarperCollins, 2010).

Windfall: The Booming Business of Global Warming by McKenzie Funk (New York: Penguin Press, 2014).

CHRONOLOGY

1769: James Watt patents his steam engine.
Atmospheric CO_2 levels are ~ 280 parts per million.

1859: John Tyndall builds the world's first ratio spectrophotometer and tests the absorptive properties of atmospheric gases.

1895: Svante Arrhenius completes his calculations on varying CO_2 levels.
Atmospheric CO_2 levels are ~ 290 parts per million.

1928: CFCs are invented.

1958: CO_2 measuring equipment is installed at the Mauna Loa Observatory.

1959: CO_2 levels stand at 315 parts per million.

1970: Paul Crutzen warns that human actions may damage ozone layer.

1979: The National Academy of Sciences issues its first major report on global warming: "We may not be given a warning until the CO_2 loading is such than an appreciable climate change is inevitable."
CO_2 levels reach 337 parts per million.

1987: The Montreal Protocol is adopted; phaseout of CFCs begins.

1988: The Intergovernmental Panel on Climate Change is established by the World Meteorological Organization and the United Nations Environment Programme.

1992: President George H. W. Bush signs the U.N. Framework Convention on Climate Change in Rio de Janeiro.
The U.S. Senate approves the Framework Convention by unanimous consent.
CO_2 levels reach 356 parts per million.

1995: The Intergovernmental Panel on Climate Change issues its Second Assessment Report: "The balance of evidence suggests a discernible human influence on global climate."

1997: The Kyoto Protocol is drafted.

1998: Average global temperatures for the year are the warmest on record.

2000: Presidential candidate George W. Bush calls global warming an "issue that we need to take very seriously."
CO_2 levels are measured at 369 parts per million.

2001: The IPCC issues its Third Assessment Report: "Most of the warming observed over the last fifty years is attributable to human activities."
A report by the National Research Council requested by President Bush states, "Greenhouse gases are accumulating in Earth's atmosphere as a result of human activities, causing surface air temperatures and subsurface ocean temperatures to rise. Temperatures are, in fact, rising."
President Bush announces that the United States is withdrawing from the Kyoto Protocol.
Third warmest year on record.

2002: Larsen B ice shelf collapses.
Third warmest year on record.

2003: Senator James Inhofe, chairman of the Committee on Environment and Public Works, says he has "compelling evidence that catastrophic global warming is a hoax."
The American Geophysical Union issues a consensus

statement asserting: "Natural influences cannot explain the rapid increase in global near-surface temperatures." CO_2 levels reach 375 parts per million.
Fourth warmest year on record.

2004: Kyoto Protocol is ratified by Russia.

2005: Extent of melt on the Greenland ice sheet reaches a record maximum.
Arctic sea ice reaches a record minimum; researchers warn sea could be ice-free in summer "well before the end of this century."
Kyoto Protocol goes into effect.
The National Academies of Sciences of the eight major industrialized nations issue a joint statement: "The scientific understanding of climate change is now sufficiently clear to justify nations taking prompt action."
The Atlantic hurricane season sets a record for the number of Category 5 Storms
Average global temperatures for the year set a new record.

2006: CO_2 levels reach 381 parts per million. Annual rise is a near-record 2.53 parts per million.
Researchers report that since 1996, the loss of ice from Greenland has doubled.

2007: The IPCC issues its Fourth Assessment Report. It states that "warming of the climate system is unequivocal" and that "most of the observed increase in global average temperatures since the mid-20th century is very likely due to the observed increase in anthropogenic greenhouse gas concentrations."
Arctic sea ice reaches a new record minimum, nearly 25 percent lower than 2005.
The U.S. Supreme Court decides that the Environmental Protection Agency has the power to regulate CO_2 under the Clean Air Act.
Al Gore and the members of the IPCC are awarded the Nobel Peace Prize.

2008: CO_2 levels reach 385 parts per million.
The Bush administration declines to issue regulations governing carbon dioxide.
Barack Obama is elected president; he states that climate change, "if left unchecked," could result in "irreversible catastrophe."

2009: International climate talks end in Copenhagen with no agreement on a successor to the Kyoto Protocol; under the nonbinding Copenhagen Accord, nations agree to try to limit warming to two degrees Celsius.
U.S. House of Representatives approves a bill to limit emissions through a "cap and trade" system.

2010: "Cap and trade" bill dies in the Senate.

2012: Superstorm Sandy hits the East Coast, causing an estimated sixty billion dollars in damages.
Artic sea ice hits another record low.
Average global temperatures tied with 2005.

2013: CO_2 levels hit 400 parts per million.

2014: NASA scientists warn that West Antarctic Ice Sheet has begun to irreversibly melt.
IPCC issues its Fifth Assessment Report, which warns that many species "will not be able to move fast enough during the 21st century to track suitable climates."
Environmental Protection Agency proposes regulations limiting CO_2 emissions from power plants.

ACKNOWLEDGMENTS

MANY VERY BUSY people gave generously of their time and their expertise to make this book possible. A number of them have been named in the preceding pages, but a number have not.

I'd like to thank Tony Weyiouanna, Vladimir Romanovsky, Glenn Juday, Larry Hinzman, Terry Chapin, Donald Perovich, Jacqueline Richter-Menge, John Weatherly, Gunter Weller, Deborah Williams, Konrad Steffen, Russell Huff, Nicolas Cullen, Jay Zwally, Oddur Sigurdsson, and Robert Correll for the help they provided on the chapters concerning the Arctic.

Similarly, I am indebted to Chris Thomas, Jane Hill, William Bradshaw, and Christina Holzapfel for their explication of evolutionary biology; to James Hansen, David Rind, Gavin Schmidt, and Drew Shindell for their lessons in climate modeling; and to Harvey Weiss and Peter deMenocal for sharing their work on ancient civilizations. Pieter van Geel, Pier Vellinga, Wim van der Weegen, Chris Zevenbergen, Dick van Gooswilligen, Jos Hermsen, Hendrik Dek, and Eelke Turkstra were extremely gracious to me when I visited the Netherlands. Robert Socolow, Stephen Pacala, Marty Hoffert, David Hawkins, Barbara Finamore, and Jingjing Qian spent many hours with me discussing mitigation strategies, while Senator John McCain, former vice president Al Gore, Annie Petsonk, James Mahoney, and Under Secretary of State Paula Dobriansky helped me to understand the politics of global warming. Mayor Pete Clavelle kindly showed me around Burlington. Michael Oppenheimer, Richard Alley, Daniel Schrag, and Andrew Weaver were always willing—and able—to answer one last question.

This book began as a series of pieces that appeared in the *New Yorker.* I am deeply grateful to David Remnick for urging—indeed compelling—me to write those pieces. I also want to thank Dorothy Wickenden and John Bennet who offered much valuable advice and encouragement along the way. I also want to thank Gillian Blake, Kathy Robbins, and Kathy Belden for all of their help, guidance, and good spirits.

The three chapters that I have added to this edition also began as stories in the *New Yorker.* Small bits of this edition also appeared in *Audubon* magazine and in my most recent book, *The Sixth Extinction.*

For their help with subjects ranging from ocean acidification to how to live on two thousand watts, I'd like to thank Victoria Fabry, Ken Caldeira, Chris Langdon, Richard Feely, Chris Sabine, James Zachos, Ulf Riebesell, Joanie Kleypas, Søren Hermansen, and Roland Stulz.

Finally, I want to thank my husband, John Kleiner, who helped in more ways than he should have. Without his peculiar optimism, not a word here would have been written.

SELECTED BIBLIOGRAPHY
AND NOTES

Most of the information contained in this book either comes from interviews or is part of the general—and vast—climate science literature. I have also cited or relied on a number of individual reports, articles, and earlier books, some of which are listed below.

Chapter 1: Shishmaref, Alaska

A study commissioned by the U.S. Army Corps of Engineers, "Shishmaref Relocation and Collocation Study: Preliminary Costs of Alternatives," December 2004, provides detailed information on the village's proposed move.

The official title of the Charney Report is "Report of an Ad Hoc Study Group on Carbon Dioxide and Climate: A Scientific Assessment to the National Academy of Sciences" (Washington, D.C.: National Academy of Sciences, 1979).

Global temperature data for the last two thousand years are drawn from Michael E. Mann and Philip D. Jones, "Global Surface Temperatures over the Past Two Millennia," *Geophysical Research Letters*, vol. 30, no. 15 (2003).

Figures on methane releases from the Stordalen mire are taken from Torben R. Christensen et al., "Thawing Sub-Arctic

Permafrost: Effects on Vegetation and Methane Emissions," *Geophysical Research Letters*, vol. 31, no. 4 (2004).
An account of the mission of the *Des Groseilliers* can be found in D. K. Perovich et al., "Year on Ice Gives Climate Insights," *Eos* (Transactions, American Geophysical Union), vol. 80, no. 481 (1999).
Figures on the thinning of the Arctic sea ice come from D. A. Rothrock et al., "Thinning of the Arctic Sea-Ice Cover," *Geophysical Research Letters*, vol. 26, no. 23 (1999).
A fuller discussion of the orbital changes and timing of ice ages can be read in John Imbrie and Katherine Palmer Imbrie, *Ice Ages: Solving the Mystery*, revised edition (Cambridge, MA: Harvard University Press, 1986).

Chapter 2: A Warmer Sky

A useful primer on the science of global warming is John Houghton, *Global Warming: The Complete Briefing*, third edition (Cambridge: Cambridge University Press, 2004).
The history of the science of global warming is related in Spencer R. Weart, *The Discovery of Global Warming* (Cambridge, MA: Harvard University Press, 2003); and Gale E. Christianson, *Greenhouse: The 200-Year Story of Global Warming* (New York: Walker and Company, 1999). Also the Tyndall Centre for Climate Change Research offers a detailed biography of its namesake on its Web site, http://www.tyndall.ac.uk.
John Tyndall's dying words, as recalled by his wife, are recounted in Mark Bowen, *Thin Ice* (New York: Henry Holt, 2005).
Svante Arrhenius's predictions for better living through CO_2 are from *Worlds in the Making: The Evolution of the Universe* (New York: Harper, 1908).
Charles David Keeling wrote about "having fun" trying to measure CO_2 in his essay "Rewards and Penalties of Monitoring the Earth," *Annual Review of Energy and the Environment*, vol. 23 (1998).

Chapter 3: Under the Glacier

An excellent account of what's been learned from the Greenland ice is Richard B. Alley, *The Two-Mile Time Machine: Ice Cores, Abrupt Climate Change, and Our Future* (Princeton: Princeton University Press, 2000).

Figures on the acceleration of the Greenland ice sheet are from H. Jay Zwally et al., "Surface Melt–Induced Acceleration of Greenland Ice-Sheet Flow," *Science*, vol. 297 (2002).

Figures on the acceleration of the Jakobshavn Isbrae can be found in W. Abdalati et al., "Large Fluctuations in Speed on Greenland's Jakobshavn Isbrae Glacier," *Nature*, vol. 432 (2004).

James E. Hansen wrote about the future of the Greenland ice sheet in his essay "A Slippery Slope: How Much Global Warming Constitutes 'Dangerous Anthropogenic Interference'?" *Climatic Change*, vol. 68 (2005).

A thorough discussion of abrupt climate change is *Abrupt Climate Change: Inevitable Surprises*, National Research Council Committee on Abrupt Climate Change, Washington, D.C.: National Academies Press (2002).

The quote from Wallace Broecker is drawn from his article "Thermohaline Circulation, the Achilles' Heel of Our Climate System: Will Man-Made CO_2 Upset the Current Balance?" *Science*, vol. 278 (1997).

The effects of the Little Ice Age in Iceland are described in H. H. Lamb, *Climate, History and the Modern World*, second edition (New York: Routledge, 1995).

The voluminous findings of the Arctic Climate Impact Assessment have been summarized in *Impacts of a Warming Arctic: Arctic Climate Impact Assessment* (Cambridge: Cambridge University Press, 2004).

Chapter 4: The Butterfly and the Toad

The most complete and up-to-date source on the habits and whereabouts of British butterflies is Jim Asher et al., *The*

Millennium Atlas of Butterflies in Britain and Ireland (Oxford: Oxford University Press, 2001).
The Victorians' passion for butterflies is documented in Michael A. Salmon, *The Aurelian Legacy: British Butterflies and Their Collectors* (Berkeley: University of California Press, 2000).
The range changes of European butterflies are described in Camille Parmesan et al., "Poleward Shifts in Geographical Ranges of Butterfly Species Associated with Regional Warming," *Nature*, vol. 399 (1999).
Information on the mating habits of frogs in upstate New York is drawn from J. Gibbs and A. Breisch, "Climate Warming and Calling Phenology of Frogs near Ithaca, New York, 1900–1999," *Conservation Biology*, vol. 15 (2001); on flowering times at the Arnold Arboretum from Daniel Primack et al., "Herbarium Specimens Demonstrate Earlier Flowering Times in Response to Warming in Boston," *American Journal of Botany*, vol. 91 (2004); on Costa Rican birds from J. Alan Pounds et al., "Biological Response to Climate Change on a Tropical Mountain," *Nature*, vol. 398 (1999); on Alpine plants from Georg Grabherr et al., "Climate Effects on Mountain Plants," *Nature*, vol. 368 (1994); and on the Edith's Checkerspot butterfly from Camille Parmesan, "Climate and Species Range," *Nature*, vol. 382 (1996).
A useful book on the biological impacts of warming is Thomas E. Lovejoy and Lee Hannah, editors, *Climate Change and Biodiversity* (New Haven: Yale University Press, 2005).
William Bradshaw published his study of *Wyeomyia smithii* living at different elevations in *Nature*, vol. 262 (1976). The evolutionary effects of climate change are documented in William E. Bradshaw and Christina M. Holzapfel, "Genetic Shift in Photoperiod Response Correlated with Global Warming," *Proceedings of the National Academy of Sciences*, vol. 98 (2001).
Jay Savage's description of discovering the golden toad, as well as a thorough description of the biology of Monteverde, can be found in Nalini M. Nadkarni and Nathaniel T. Wheelwright, editors, *Monteverde: Ecology and Conservation of a Tropical Cloud Forest* (New York: Oxford University Press, 2000). Details of

the golden toad's life cycle are drawn from Jay M. Savage, *The Amphibians and Reptiles of Costa Rica: A Herpetofauna Between Two Continents, Between Two Seas* (Chicago: University of Chicago Press, 2002).

The demise of the golden toad is linked to precipitation patterns in J. Alan Pounds et al., "Biological Response to Climate Change on a Tropical Mountain," *Nature*, vol. 398 (1999). Pounds's findings have since been widely challenged, and it now seems more likely that the golden toad was driven to extinction by a pathogen, the fungus *Batrachochytrium dendrobatidis (Bd)*. See Karen Lips et al., "Riding the Wave: Reconciling the Roles of Disease and Climate Change in Amphibian Declines," *PLOS Biology*, March 25, 2008. Efforts to model the future of the cloud forest are detailed in Christopher J. Still et al., "Simulating the Effects of Climate Change on Tropical Montane Cloud Forests," *Nature*, vol. 398 (1999).

The quotes from G. Russell Coope are drawn from his essay "The Palaeoclimatological Significance of Late Cenozoic Coleoptera: Familiar Species in Very Unfamiliar Circumstances," which appeared in Stephen J. Culver and Peter F. Rawson, editors, *Biotic Response to Global Change: The Last 145 Million Years* (Cambridge: Cambridge University Press, 2000).

The figures on potential extinctions are drawn from C. D. Thomas et al., "Extinction Risk from Climate Change," *Nature*, vol. 427 (2004).

Chapter 5: The Curse of Akkad

An introduction to Akkadian civilization can be found in Marc Van De Mieroop, *A History of the Ancient Near East* (Malden, MA: Blackwell Publishing, 2004).

The verses from *The Curse of Akkad* come from Jerrold S. Cooper, *The Curse of Agade* (Baltimore: The Johns Hopkins University Press, 1983).

A detailed description of Tell Leilan can be found in the chapter on the site by Harvey Weiss in *The Oxford Encyclopedia of*

Archaeology in the Near East, vol. 3, Eric M. Meyers, editor (Oxford: Oxford University Press, 1997). Climate change was first proposed as the cause of the abandonment of Tell Leilan in Harvey Weiss et al., "The Genesis and Collapse of Third Millennium North Mesopotamian Civilization," *Science,* vol. 261 (1993).

An overview of the connections between climate change and societal collapse can be found in Peter B. deMenocal, "Cultural Responses to Climate Change During the Late Holocene," *Science,* vol. 292 (2001).

The findings of American scientists in Lake Chichancanab are described in David Hodell et al., "Possible Role of Climate in the Collapse of Classic Maya Civilization," *Nature,* vol. 375 (1995). The findings of researchers off the coast of Venezuela are described in Gerald Haug et al., "Climate and the Collapse of Mayan Civilization," *Science,* vol. 299 (2003). A detailed discussion of drought and Mayan civilization can be found in Richardson B. Gill, *The Great Maya Droughts: Water, Life, and Death* (Albuquerque: University of New Mexico Press, 2001).

James Hansen spoke of being "captivated" by the study of greenhouse warming in a speech titled "Dangerous Anthropogenic Interference: A Discussion of Humanity's Faustian Climate Bargain and the Payments Coming Due," delivered at the University of Iowa, October 26, 2004.

Predictions of warming-induced water shortages in the United States come from David Rind et al., "Potential Evapotranspiration and the Likelihood of Future Drought," *Journal of Geophysical Research,* vol. 95 (1990).

Peter deMenocal wrote about the connections between climate change and human evolution in "African Climate Change and Faunal Evolution During the Pliocene-Pleistocene," *Earth and Planetary Science Letters,* vol. 220 (2004).

Evidence of the drought in Tell Leilan in sediments from the Gulf of Oman is presented in Heidi Cullen et al., "Climate Change and Collapse of the Akkadian Empire: Evidence from the Deep Sea," *Geology,* vol. 28 (2000).

The connection between climate change and the disintegration of

Harappan society is discussed in M. Staubwasser et al., "Climate Change at the 4.2 Ka BP Termination of the Indus Valley Civilization and Holocene South Asian Monsoon Variability," *Geophysical Research Letters*, vol. 30, no. 8 (2003).

Chapter 6: Floating Houses

Figures on the Netherlands' water management system come from the Dutch Ministry of Transport, Public Works and Water Management, *Water in the Netherlands: 2004–2005* (The Hague, 2004).

Figures on sea level rise are drawn from the Intergovernmental Panel on Climate Change, *Climate Change 2001: The Scientific Basis*, J. T. Houghton et al., editors (Cambridge: Cambridge University Press, 2001).

The study of flooding commissioned by the British government is cited in David A. King, "Climate Change Science: Adapt, Mitigate, or Ignore?" *Science*, vol. 303 (2004).

A detailed analysis of the Vostok core can be found in Jean Robert Petit et al., "Climate and Atmospheric History of the Past 420,000 Years from the Vostok Ice Core, Antarctica," *Nature*, vol. 399 (1999).

Discussions of the threshold for "dangerous anthropogenic interference" can be found in Brian C. O'Neill and Michael Oppenheimer, "Dangerous Climate Impacts and the Kyoto Protocol, *Science*, vol. 296 (2002); and James Hansen, "A Slippery Slope: How Much Global Warming Constitutes 'Dangerous Anthropogenic Interference'?" *Climatic Change*, vol. 68 (2005).

Chapter 7: Business as Usual

The Environmental Protection Agency's personal emissions calculator can be found at: http://yosemite.epa.gov/oar/globalwarming.nsf/content/ResourceCenterToolsGHGCalculator.html.

Stephen Pacala and Robert Socolow laid out their "wedge" scheme in "Stabilization Wedges: Solving the Climate Problem for the Next 50 Years with Current Technologies," *Science*, vol. 305 (2004).

Information on the fuel efficiency of U.S. autos comes from the report "Light-Duty Automotive Technology and Fuel Economy Trends," Advanced Technology Division, Office of Transportation and Air Quality, U.S. Environmental Protection Agency, July 2005.

The need for new energy technologies to stabilize CO_2 is discussed in Martin Hoffert et al., "Advanced Technology Paths to Global Climate Stability: Energy for a Greenhouse Planet," *Science*, vol. 298 (2002); and also in Hoffert et al., "Energy Implications of Future Stabilization of Atmospheric CO_2 Content," *Nature*, vol. 395 (1998). Martin Hoffert and Seth Potter wrote about space-based solar power in "Beam It Down: How the New Satellites Can Power the World," *Technology Review*, October 1, 1997.

Chapter 8: The Day After Kyoto

Former treasury secretary Paul O'Neill's speculations about Vice President Dick Cheney are recounted in Ron Suskind, *The Price of Loyalty: George W. Bush, the White House, and the Education of Paul O'Neill* (New York: Simon & Schuster, 2004).

The high degree of consensus on global warming is documented by Naomi Oreskes, "The Scientific Consensus on Climate Change," *Science*, vol. 306 (2004).

The Bush administration's editing of climate science reports was disclosed by Andrew C. Revkin, "Bush Aide Edited Climate Reports," *New York Times*, June 8, 2005.

The Bush administration's efforts to water down the proposal for joint action at the 2005 G8 Summit were reported by Juliet Eilperin, "U.S. Pressure Weakens G8 Climate Plan," *Washington Post*, June 17, 2005.

Chapter 10: Man in the Anthropocene

Paul J. Crutzen wrote about the dawn of the Anthropocene and the "luck" that spared the world from catastrophic ozone loss in his essay "Geology of Mankind," *Nature*, vol. 415 (2002). Sherwood Rowland recounts his reaction to his discovery in Heather Newbold, editor, *Life Stories: World-Renowned Scientists Reflect on Their Lives and the Future of Life on Earth* (Berkeley: University of California Press, 2000). The discovery of the "ozone hole" is related in Stephen O. Anderson and K. Madhava Sarma, *Protecting the Ozone Layer: The United Nations History* (London/Sterling, VA: Earthscan Publications, 2002). The amount of warming still required to bring the earth into energy balance is discussed in James Hansen et al., "Earth's Energy Imbalance: Confirmation and Implications," *Science*, vol. 308 (2005).

Chapter 11

The two studies that point to irreversible melt of the Amundsen Sea sector are: E. Rignot et al., "Widespread, Rapid Grounding Line Retreat of Pine Island, Thwaites, Smith, and Kohler glaciers, West Antarctica, from 1992 to 2011," *Geophysical Research Letters*, vol. 41 (2014); and Ian Joughin et al., "Marine Ice Sheet Collapse Potentially Under Way for the Thwaites Glacier Basin, West Antarctica," *Science*, vol. 344 (2014). Changes to *National Geographic's Atlas of the World* were reported in Christine Dell'Amore, "Shrinking Arctic Ice Prompts Drastic Change in National Geographic Atlas," nationalgeographic.com, published online June 9, 2014. Data on Americans' priorities from Pew Research, "Thirteen Years of the Public's Top Priorities," published online January 27, 2014, http://www.people-press.org/interactives/top-priorities/ For more discussion of the misinformation campaign, see Naomi

Oreskes and Erik M. Conway, *Merchants of Doubt: How a Handful of Scientists Obscured the Truth on Issues from Tobacco Smoke to Global Warming* (New York: Bloomsbury Press, 2010).
J.E.N. Veron's gloomy conclusions come from his article, "Is the End in Sight for the World's Coral Reefs?" *Yale Environment 360*, published online December 2010, http://e360.yale.edu/feature/is_the_end_in_sight_for_the_worlds_coral_reefs/2347/.
Bill McKibben's "Global Warming's Terrifying New Math" was published in *Rolling Stone*, July 19, 2012.

Chapter 12: The Darkening Sea

The Royal Society's report on ocean acidification is "Ocean Acidification due to Increasing Atmospheric Carbon Dioxide," published June 30, 2005, and available online at https://royalsociety.org/policy/publications/2005/ocean-acidification/.
Ken Caldeira's paper modeling the effects of carbon emissions on the oceans is "Anthropogenic Carbon and Ocean pH," *Nature*, vol. 425 (2003).
Chris Langdon's first paper on the corals of Biosphere 2 was "Effect of Calcium Carbonate Saturation State on the Calcification Rate of an Experimental Coral Reef," *Global Biogeochemical Cycles*, vol. 14 (2000).

Chapter 13: Unconventional Crude

Alex Farrell and Adam Brandt's calculations about the shift to unconventional fuels can be found in "Scraping the Bottom of the Barrel: Greenhouse Gas Emission Consequences of a Transition to Low-Quality and Synthetic Petroleum Resources," *Climatic Change*, vol. 84 (2007).

Chapter 14: The Island in the Wind

The 2000-Watt Society's "white book" is available online at http://www.novatlantis.ch/fileadmin/downloads/2000watt/A%20White%20Book%20on%20RandD.pdf

INDEX

Note: Information presented in figures is denoted by *f*.

A NOTE ON THE AUTHOR

Elizabeth Kolbert is a staff writer for the *New Yorker*. She is a two-time National Magazine Award winner and has also received a Heinz Award and a Guggenheim Fellowship. Kolbert lives in Williamstown, Massachusetts, with her husband and three sons.

A NOTE ON THE TYPE

The text of this book is set in Bembo. This type was first used in 1495 by the Venetian printer Aldus Manutius for Cardinal Bembo's *De Aetna*, and was cut for Manutius by Francesco Griffo. It was one of the types used by Claude Garamond (1480–1561) as a model for his Romain de L'Université, and so it was the forerunner of what became standard European type for the following two centuries. Its modern form follows the original types and was designed for Monotype in 1929.